Andrew Jackson Davis

The Harbinger of Health

Containing medical prescriptions for the human body and mind

Andrew Jackson Davis

The Harbinger of Health
Containing medical prescriptions for the human body and mind

ISBN/EAN: 9783744796538

Printed in Europe, USA, Canada, Australia, Japan

Cover: Foto ©berggeist007 / pixelio.de

More available books at **www.hansebooks.com**

HARBINGER OF HEALTH;

CONTAINING

MEDICAL PRESCRIPTIONS

FOR THE

HUMAN BODY AND MIND.

BY

ANDREW JACKSON DAVIS.

Complete in One Volume.

Health is the first condition of Happiness.
The Angels help those who help themselves.

NINTH THOUSAND.

BOSTON:
PUBLISHED BY BELA MARSH,
14 Bromfield Street.
1868.

PREFATORY QUESTIONINGS.

The author would confer a few moments with every one who proposes to take refuge in this HARBINGER OF HEALTH The present work is, perhaps, a vestibule to some new temple of life, health, and happiness. It is, however, deemed judicious to ask and answer a few practical questions in the presence of the reader, so that the aims and objects of this volume may be fully understood:

QUESTION. What is the chief end of the earthly life of man?

ANSWER. To individualize his spirit, and prepare it for the Summer Land.

Q. What is the first condition of such individualization and preparation?

A. Physical health.

Q. In what does physical health consist?

A. In symmetry of development, energy of Will, harmony of function, and bodily purity.

Q. How can these results be obtained?

A. First, by inheriting a sound constitution; second, by obeying the law of temperance in regard to foods and drinks; third, by giving free play and equal exercise to the muscular system; fourth, by exerting the Will-power to keep the passions in subjection; fifth and lastly, by sleeping and working and living in accordance with the requirements of Nature's Laws.

Q. What are the penalties of disobedience?

A. Angular development, physical weakness, pains, aches, discord, decrepitude, and disgust.

Q. What is this condition called?

A. It is called Disease.

Q. What are the physical and mental consequences of Disease?

A. Bodily unhappiness and mental misery.

Q In what does Disease consist?

A. In an excess, misplacement, or deficiency.

Q. How can an excess be overcome?

A. By supplying the deficiency of some unfulfilled want.

Q. What is the first condition of restoration?

A. Purification of both the skin and the digestive system.

Q. How can these ends be accomplished?

A. By thoroughly washing the body and lightly anointing the entire cuticle; the second result may be obtained by fasting for several days.

Q What is the *immediate* cause of Disease?

A. Exhaustion of the vital forces (*i. e.*, the spiritual dynamics of the soul;) also excessive impurity, in some particular organ or fluid.

Q. What are the most prominent causes of such exhaustion and impurity?

A. The causes in this country are, first, badly prepared food and rapid eating; second, deficient or excessive exercise; third, uncleanness of the skin; lastly, unhealthy respiration.

Q. What is the immediate cause of impurity?

A. The retention of inappropriate matter in the system.

Q. Why is such foul matter retained?

A. Because of the ignorance or negligence of man.

Q. What are the proper channels for the expulsion of foul matter?

A. The channels are six: The skin, the nerves, the lungs, the liver, the kidneys, and the bowels.

Q. What is the process of expelling the impure matter which produces Disease?

A. The process is called excretory, because the blood sends off its feverish exhalations, in the form of *perspiration*, through the skin; the mind sends off its surplus magnetism and electricity, in the form of *activity*, through the nerves; the blood-making organism sends off its poisonous smoke, in the form of *carbonic acid*, through the lungs; the fluids of the entire body discharge their most acrid portion, in the form of *urine*, through the kidneys; the entire organ-building mechanism expels its broken-down material, its cinders and ashes, in the form of *bile*, from the liver; and lastly, the inappropriate and refuse matter of the whole body is thrown off through the intestines. It therefore follows that, if any one of these excretory processes be neglected, or obstructed in any degree, the consequences are physical derangements and some form of acute Disease.

Q. What is an acute Disease?

A. It is a sudden and violent effort of Nature to expel impurities from the system.

Q. What is a chronic Disease?

A. It is a less violent and more tedious effort to restore the conditions of health.

Q. What are the sources of nervous power and energy?

A. Healthy diet, pure air, ample exercise, and a good digestion.

Q. What conditions are necessary to a good digestion?

A. Simple food, well masticated; a clean skin, and contentment of mind.

Q. What is simple food?

A. An admixture of healthy substances, not to exceed *three* solid articles, for one meal; and a very temperate use of butter, salt, vinegar, sugar, pepper, tea, and coffee.

1*

Q. What will produce mental contentment?

A. A full and complete realization of the truths of the HARMONIAL PHILOSOPHY.

Q. In what way may we obtain pure air in a sickly district or low country?

A. By breathing habitually through the nose, with the mouth closed; at night, hang a wet sheet or blanket near the open window of the sleeping apartment.

Q. What is the effect upon the air of breathing it through the nose?

A. Its temperature is immediately raised, which acts upon impure air just as boiling acts upon impure water, thus preparing the air for the uses of the lungs. The wet sheet unloads the air of its diseasing miasm, such as carbonic acid and various animal emanations, by which the lungs are depressed and the vital system poisoned.

Q. Should the common air always enter the lungs through the nose?

A. Yes, unless while taking violent exercise, such as running and jumping, when the heat of the mouth and throat is sufficient to purify and prepare the air for the lungs. It should be remembered that the nose is to the lungs what the mouth is to the stomach. As you would not think of swallowing food *unmasticated*, so you should not permit yourself to inhale air *unpurified*. The teeth and mouth, with their juices, accomplish the one; and the nose, with its various compartments and fluids, performs the other.

Q. Besides air, what is the next most important agent of health?

A. The agent of life and health, next to air, is water.

Q. What is the evidence that water is thus important?

A. It completely pervades the earth outside of man, and forms over four-fifths of his physical constitution; it forms more than ninety per centum of his blood; is absorbed by every membrane in his body; is the element on which all the particles float from part to part; is essential to all digestion, secretion, breathing, perspiration, and purification.

Q. Is water the best medicine in a case of Disease?

A. The human body is never healed by any one agent or element. This book will impart a correct knowledge of the best general treatment for diseased conditions.

Q. Is it possible to live in this world entirely free from bodily ailments?

A. No. The most careful and obedient person will not always escape slight functional disturbances.

Q. Why is this so?

A. Because the globe is yet young, and untamed or uncultivated. Its water, soils, plants, animals, and air, are not yet sufficiently refined and purified to prevent disease.

Q. Will the time ever come on earth when Disease shall be no more?

A. Yes. And even "Death" will one day be swallowed up in victory.

Q. How will this be possible?

A. There will be no "Disease" when the globe shall have become perfectly subdued and gardenized by man; and there will be "no Death" when the earth's inhabitants shall perfectly realize the nearness of the Summer Land.

Q. What is the common object of life?

A. Happiness.

Q. Why are not more persons happy?

A. Because the mass of mankind err as to the true means of happiness.

Q. What are the true means of happiness?

A. This all-important question is earnestly analyzed, and, we trust, philosophically answered, in the four hundred and twenty pages which compose this volume. And to the vast Brotherhood of mankind, therefore, the contents of this book are fraternally dedicated by   THE AUTHOR.

NEW YORK, October 15th, 1861.

CHAPTER I.

THE PEARLY GATES OF SCIENCE.

NATURE's harmonious, eternal heart, is incessantly throbbing with the almighty energy of omnipotent principles. Sweetly bloom the progressive truths and immortal beauties of the Infinite. They come gracefully out from their invisible sanctuary, and shine steadily and lovingly into the gloomy abysses of ignorance.

As a brief definition, we may say that Science is a knowledge of Facts and Forces. What is Art, then, but the intellectual and manual power to control such forces for the gratification and benefit of mankind? There is a plain difference between Art and Science. The latter is the embodiment of intellectual discoveries; the former is the archangel which puts theory into practice, for the world's permanent good.

If Science is the glory of mind, then Art is its crown of immortality. But mankind are admonished to travel for forty years in the wilderness of facts not only, but to traverse and reverse their contemplations of the Universe forever, in quest of the countless treasures which lie within the bosom of the Summer Land. When the pilgrim arrives at the goal of Scientific Knowledge, no matter what path he may have pursued, the angels bring forth and place upon his brow a royal diadem, in these days called "COMMON SENSE!"

Such a mind feels that Facts are the temporary, yet necessary, stepping-stones of individual progress. Facts are the hard currency of the intellect. They systematize its operations, demolish its flickering superstitions, and promote the refining and useful Arts.

Those who would attend the Academy of right-thinking must take primary lessons at the feet of Scientific truths. Facts in geology, facts in chemistry, facts in physiology, facts in history, facts in mechanism, facts in spiritualism—facts, facts, from and of every side in the rolling Universe—are the first firm friends of Common Sense, and the most trustworthy intercessors between mind and matter. Genius and transcendent talents, even reason and intuition, are next to impossible, except through obedience to the gospel of Facts.

But the lesson of Facts is impressive and startling in its sublimity. This lessons is, that facts are *never* finalities! They will let you dogmatize only for a brief season. They neither begin nor terminate in and of themselves. They crop out from a hidden soil, and jut over into some invisible world. To the philosopher they are but palpable shadows thrown upon the soaring mountains of thought. They indicate the existence of invisible substances. They send divinest dispatches to every mind, saying: "Seek further, if thou wouldst behold the Principles of the Infinite."

Excelsior! is the song of every fact. Beauteous flower-facts bedeck the far-spreading prairies of vision. But the fire-chemist can banish them into the boundless ocean of imponderables. These imponderables invite you "still higher" among the impersonals. These impersonals, which only the exquisite sense of intuition can appreciate, are the Principles of Nature. But it is just as true to say that they are the Will-powers of Father God.

Thus Facts lead away into Principles, which are the rest and delight of the harmonious mind. We are, therefore, admonished to observe and follow the lead of Scientific Facts. Men should teach their children to know some facts not traditional, but which are *absolutely* and unequivocally Scientific. Otherwise the mind is chaotic, confused, uncertain, and exposed to the epidemical influence of priests and superstition. Ministers who believe only in the facts of tradition—who refuse the living facts of a living Age—are certain to take possession of minds not fortified with absolute Facts. The drifting affections of the ignorant are wafted by every wind of doctrine. A few well-ascertained and positive facts in Magnetism, for instance, have proved "an anchor to the soul, both sure and steadfast;" while a neighboring bark has been driven by the gusty breathings of a prayer-meeting upon the rocks of Bigotry and Despair.

The priest-paralyzed Galileo was not killed, because a shield of facts hung between him and his persecutors. The Alpine storms of priestcraft could not drive him ashore. His mind stood firmly upon a foundation of astronomic facts, and amid the deafening thunders of the Vatican he cried—"the world does move!" In like manner we may record the progressive development of all Sciences and Arts, as also of their masters. The burning breath of an hundred theological Saharas could not put out the eyes of Newton. Neither did the world of frowning superstition weaken the energies of Spinoza, Kant, Gall, Spurzheim, Combe, and more familiar minds, who have wrought so bravely for the advancement of common sense and mankind.

Put a few positive Facts before your child's mind, not dryly and severely, as though each fact was a "stubborn thing," fit only for headstrong boys, but gently teach your child facts and truths of the world in which we live, and thus fortify the reason ing powers against the assaults of weakness and superstition. Facts are foundation stones in the crystal palace of human knowledge. But, remember, the foundation—below the facts and temple of Progress—is composed of Eternal Principles. Facts are but the incarnations of these immortal Verities. Yet it is ordained, in the system of moral and intellectual growth, that each mind shall pass through the pearly gates of Science.

But human ignorance is demoniac darkness of the blackest degree—reflecting none of the rays of wisdom—hence it does not comprehend the sublime import of truth. To the dull-minded man, the physical world seemeth dull and dirty—no light is flowing for him from every well-spring of wisdom, in field and forest—and yet no murmurings, no evil voices, no sobbing complaints, ever break from Nature's seemingly neglected Soul. But the miserable millions of earth groan and weep—and why? Not because they have faithfully labored in the vineyards of Progress, and failed to realize great crops therefor, but because they erringly have chosen the delusive slumbers of idleness and ignorance. They have not accepted of the "golden key to ope the palace of Eternity"—have not sought to comprehend the developments of endlessly progressive principles—but have, instead, dwelt idly and willfully in the mysterious shadows of mental blindness and bigotry. Say to the world, in tones as mighty as the thundering voice of earthquakes, that "Salvation from all disease and discords—physical, social, political, and spiritual—is possible only by and through *personal* obedience to every known requirement of the law of love and justice," and how many of those who believe in salvation by and through the blood and sufferings of an ancient Martyr, would respond "Amen"? Only those would accept of such terms who wisely love truth and justice. All the idle and morally ignorant would abide with the wrongly-educated of the churches, and thus help swell the murky rivers of bigotry and regressive conservatism.

But the divine secrets of Nature will out! Herbs, shrubs, and trees, do not more naturally grow *outside* the churches than do the Arts, Sciences, and Philosophies flourish independently of ordained priests. The eternal momentum of progressive truth forces *light* through what appear to be impenetrable mountains of ignorance. As cultivation subdued the savageness and consequent *crabbedness* of the tiny wild apple, making it one of the most umbrageous growths of our orchards, and almost the finest fruit in this latitude, so will the silent workings of progressive truth unfold the sourest tempers into springs of sweetness, fertilize the germs of beauty in the ugliest bosoms, adorn with Reason the flat, low brow of ignorance or misfortune, and convert the primitive instincts of crude souls into flowering and fragrant plants of Intuition.

Yesterday we heard bitter anathemas and heart-stricken weepings, because the ruthless hand of some headstrong Reformer had demolished sacred idols and overthrown moss-covered cathedrals; but to-day, as we walked that way in deepest meditation, we heard songs of gladness and shouts of enthusiasm, because of joys unspeakable, which had taken possession of the spiritually unfolded and free! The flowing and swelling anthem written by the Poet Dana, was the grandest!

"O Listen, man!
A voice within us speaks that startling word,
Man! thou shalt never die! Celestial voices
Hymn it to our souls; according harps
By angel fingers touched, when the mild stars
Of morning sang together, sound forth still
The song of our great Immortality.
Thick clustering orbs, and this our fair domain,

> The tall dark mountains and the deep toned seas,
> Join in this solemn universal song.
> O listen ye, our spirits! drink it i
> From all the air! 'Tis in the gentle moonlight;
> 'Tis floating midst Day's setting glories; Night,
> Wrapt in her sable robe, with silent step
> Comes to our bed and breathes it in our ears.
> Night and the Dawn, bright Day and thoughtful Eve:
> All time, all bounds, the limitless expanse,
> As one vast mystic instrument, are touched
> By an unseen living hand, and conscious chords
> Quiver with joy in this great jubilee.
> ———The dying hear it, and as sounds of earth
> Grow dull and distant, wake their passing souls
> To mingle in this heavenly harmony."

The prospective triumph of knowledge over ignorance, of wisdom over folly, of Harmonial Philosophy over Discordant Theology, seems universal and complete. Substantially built temples of attractive industry, the thoroughly organized homes of happiness and distributive justice, will occupy acres now covered by massive structures of mythology and bigoted superstition! The social wilderness of vice and misery will become a landscape of life and loveliness unsurpassed. Eternal Use and its spirit bride, Beauty, will dwell in every habitation. Birds and animals, no longer fearing, will love their lord and master; and instead of the old heavens of storm-clouds, and the old earth of thorns and thistles, "all things shall be changed."

Who, let us now ask, can resist the aggressive march of truth? What power do you know of, either in Church or State, that can roll back upon Nature's heart the outward flowing tides of wisdom? Who can lacerate and cripple the beautiful feet of the Arts and Sciences? Do you fear? "O, ye of little faith!" Do you not see the majesty of truth? Surely, the churches will float upon the deluge of eternal principles that is to come! Let the Arts and Sciences flow forward; the "gates of hell" (discord and ignorance) "shall not prevail against them." Priesthoods must eventually disappear beneath the sun-flashes of justice and truth.

Truth is incessantly speaking from the inner temple of human life. Ignorance is the only misfortune and mystery; for that which is perfectly known is neither hurtful nor mysterious. If a strong man be ignorant of beauty, and in the haste of his blood, his tongue knoweth no emphatic language suitable to ears refined, he belches forth the words of profanity—even in such a mind the truth will out; so the ways of God are not "past finding out." In a fine sense, He is "as full, as perfect, in hair as heart." His ways *come out* daily, hourly, momentarily—in the bright and black, in the wise and foolish, in the ugly and beautiful, upon great mountains, and in all the valleys, in all human history, and upon every star in the bending firmament—and thus He is "All in All!"

Will not error and delusion vanish in proportion to the approach of truth and reason? It is our great happiness to believe that the inharmonious and conflicting elements of society will one day obey the commandment, "Peace, be still." This is the voice of the Eternal Spirit. It attaineth utterance through the mouth of every inspired soul, who loveth justice and harmony between man and man. The good-minded everywhere, who fill high stations in Church and State, will, in due time, join the army of the Free! *There is no freedom outside of truth.*

Absurd and vain is the attempt of bigots to stay the advancement of eternal principles. Truth will out! God is absolute Truth, and Nature is absolute Truth; and one is no more divine and eternal than the other; both are unchangeable and irresistible. Behold! ye faithless ones of earth, the world of mind is in energetic motion. Its life is throbbing and yearning for harmonious and peaceful expression. The whole heavens are teeming and beaming with spiritual intelligences. They must speak, and their voices will be heard (O, how gladly!) by the listening ear of the millions. But the very proud, the noble of earth, the so-called educated—how *slow*, how *conceited*, how *large* with emptiness, how *incapacitated* for the simple truths of wisdom!

But to reach the thronging multitudes of every country, to educate the masses in the natural ways of spiritual truth, to convert the sectarian churches, to simplify and elevate the systems and principles of government—what an easy task, indeed, when man's reason begins fully to realize the true philosophy of Nature and destiny.

We look with joy unspeakable for the triumph of Science over silliness; of Reason over ignorance, which holds its sessions of superstition in every land. When untrammeled and reverent Reason shall sweep the boundless horizon of Progress —just as the telescope describes the measureless abysses of starry worlds—then will astounding discoveries be made in all the kingdoms of Nature. Labor-saving machines will bring at once relief to muscle and growth to mind. Then will mankind's earthly condition, both physical and mental, be ameliorated and improved. And the paths of progress will not lead, as now, through gloomy forests of error and bloodshed; but, instead, each soul will be attracted righteously by the celestial rays of eternal Truth. "Truth is mighty, and will prevail!"

CHAPTER II.

THE PHILOSOPHY OF DISEASE.

The Philosophy of Disease is given in the first volume of the Harmonial series. We are not fond of re-writing our positions on any subject; yet it may be profitable to make a new statement of a momentous question. Judging from the number and singularity of the letters we have recently received, a large proportion from several well-educated physicians, we conclude that, even at this late day, the Harmonial Philosophy of disease is not very well understood. For the purpose of further explaining our positions, and particularly to erect a standard, or standpoint, from which to analyze the different theories of popular schools, we proceed to define our philosophy of disease. We would invite the candid-minded, whether physicians or not, to give the following statements more than a passing perusal.

The human body is constructed *double* throughout. Binary compositions, both solid and fluid, are everywhere visible. The right and the left, the perfect adaptation of structures exactly opposite and dissimilar in function, are facts within every man's observation.

Now, let it be observed, each operation within the human body occurs upon the same general plan. Reciprocal actions are perfectly exhibited in every part. The reciprocal operations are attended with *periodical* changes. These periodical changes are attended with either an *increased* or else a *diminished* degree of temperature. This temperature exerts one of two influences: first, it either agrees with and promotes healthy conditions, or it does not. If the bodily temperature does not accord with that balanced state which we term "Health," the result is a change among the atomic motions of the parts most affected by the disturbance; and the next, or succeeding effect, is either too much or too little heat (or vital fire) in the assailed organs or functions. If too much heat exists in a particular part, then the atoms of that part are in too rapid motion for health, and the first development is an "inflammation." The part or parts are "on fire;" while in some part or parts, the atomic motions are at the same time too slow, to a corresponding extent, and the result in those localities is the development of a "chill"—all because those departments are deprived of their appropriate quantity of the vital fire, which is the *soul principle*, or rather the ultimate organization of the immortal and far more interior spirit.

Extreme   Extreme

CHILL.——HEALTH.——FEVER.

Negative.   Positive.

Of course, as you readily comprehend, no substance is moved without being first actuated by some adequate principle or force. Wherefore let the doctrine be established that, whenever the body is disturbed from its center of gravity, whenever it is made to wabble in its orbit, or caused to vibrate in the sphere of life, the producing causative power may be traced primarily to that soul-principle, which animates the physical temple, and which clothes the mind, spirit, or understanding—in shortest word, the whole interior individualized MAN.

I. ORIGIN OF DISEASE.

All diseases, therefore, are referable, as to their origin, to the soul principle. The moving principle is first disturbed; then, in the lower scale, the material parts take on a corresponding disturbance. Suppose, for example, you absorb a dose of arsenic. This mineral is a deadly poison. But would it poison a stone jug? Put the portion of poison down a leaden pipe, instead of consigning it to your stomach, would the pipe experience death by arsenic? The question answers itself. But why is the pipe not poisoned? Because, we reply, the lead is not animated by a *principle* that can be thus disturbed. Man's dead body could not be poisoned by calomel, nor awakened by the thunders of a nation's artillery, because the soul-principle is not within it.

Matter will not respond intelligently to any appeal, neither will it, separated from the vital intelligence, generate either pain or pleasure. Therefore, it is the hight of absurdity to expect physical results to follow the lead of mere material bodies. Bodies without souls are dead. If you drink a glass of brandy, the soul-principle is first disturbed by the spirit within the fluid; then, as a natural and inevitable sequence, the body itself, with its varied parts and functions, receives a record of the disturbance; and the magnitude of such disturbance is always proportional to the extent to which the soul-principle is thrown from its accustomed equilibrium. A cup of tea or of coffee, a dinner of beefsteak, and the like, exert each a direct influence on the animating principle to begin with; and then the condition of the physical organs and functions, subsequently, will ever be an exact index of the *kind* and *extent* of influence that was generated by and within the principle.

II. PERIODICITY OF DISEASE.

By virtue and in pursuance of the soul-principle, which is composed of positive and negative attributes, all disturbances have *ups* and *downs*, or seasons of exacerbation and severity, followed by corresponding periods of mildness and diminution. There is a perfect system of *vibration* between two opposite points, or rather a swinging of the *health-pendulum* from one extreme to the other; and the number of changes in the atoms of the body may be mathematically determined by the velocity of the vibration. Suppose, for illustration, you drink a bowl of strong coffee. The soul-principle will very rapidly respond to the spirit within the beverage. Without detailing the chemical facts in the premises, we will mark the effect upon the ganglionic centers, and in the throbbings of the heart. The ganglia are larger, and the pulse beats faster than before you imbibe the coffee. This condition is what we term the exacerbation—the hight of the action—the maximum state of the disturbance. But reaction will, sooner or later, succeed to a corresponding degree; when the condition of the body will fall as far below the sphere of Health as the previous physiological operations were exalted too far above it. Both disturbances are diseases, or rather they tend to establish and promote those fictitious soul-alterations which invalidate the integrity of the vital energies, and thereby generate the conditions of disease.

The importance of this *periodicity in disease* is not likely to be apprehended without due reflection. Take experience for your teacher, try this philosophy by its lessons, and see whether we do not report Nature faithfully.

Say, to begin with, you are in a perfect state of bodily health. You do not think of your body, either while resting or working. It is without pain and fatigue. The river of energy flows over rocks or along grassy margins, all the same without awakening bodily ills, or allaying pains you do not possess. But your system is perfectly and mathematically *periodical* during all this life of health. Observe; Are you not *sleepy* at a particular hour of the evening, and do you not *arouse* from slumber at a particular hour on the subsequent morning? Does not your healthy appetite *call for food*, day by day, at a certain never-varying notch in the sun-dial? All the natural functions go forward with the beautiful regularity and precision of the completest clock-work. If you till the soil, there is a particular portion of the day which contributes more to your labors than all other portions. You can work more, easier, and better, at some special section of the diurnal circle. If this time came on yesterday at 11 o'clock, then to-day at that hour, you will feel similarly inclined and energetic. Or, suppose you should retire from your work at 10 o'clock to-day, and then engage in quiet conversation till dinner-time; the chances are that, if you watch your sensations, you will feel disinclined to continue your labor at precisely the same hour to-morrow. A close observer will fully indorse these affirmations.

Again: Suppose you indulged last night in some delightful amusement, the excess of which brings weariness upon you through all the early hours of to-day, the natural effect will nevertheless be to cause you to realize an undefinable wish for a repetition of the same indulgence to-night at about the same hour. On this principle of *periodicity*, you will observe strong-minded men giving themselves in utter abandon to the wine cup, to reckless intemperance, and voluptuous debauchery, at particular parts of the year or month, while the same men are examples of sobriety and good citizenship at almost all other periods. Children wish to play periodically; that is, at the same hour every day. Artists are inspired at some marked portion of the twenty-four hours. Each mind chooses its own period, either day or night, and is influenced accordingly.

Nature is thus full of physiological and psychological *antitheses*. She brings the rhetoric of Health before the imagination of her children in no other way. It is natural for rhetoricians to point an argument with a brilliant antithesis; such as *aversion*, as opposed to *desire ;* *past*, as the extreme other end of *future ;* *identity*, as the opposite of *contrariety; truth*, as the antagonist of *error ;* and so, in physiology, we use *inspiration* and *expiration*, *endosmosis* and *exosmosis*, *secretion* and *excretion*, *action* and *repose*, *chill* and *fever*, because it is the instinct of reason to look for the counterpart, or for *the other end*, of that which is felt or seen.

III. THE TREATMENT OF DISEASE.

In treating with disease, two things are to be kept in mind: first, the fact that all disturbances of the physical structure originate in the soul-principle; and second, that the laws of equilibrium make it certain that all disturbances shall pass *periodically* from one extreme to the other.

Medicines, therefore, are appropriate only when they meet the wants of the disturbed principle—*at the period* when it is struggling hardest to restore the material parts to their own places in the temple. To give medicine or magnetism when the soul-principle is comparatively quiet—that is, when it is about midway between the two extremes—is exceedingly false to the requirements of either body or soul. For example: Suppose you ascertained last night, for the first, that your lungs are covered with "a cold." That is a sufficient proof that you will feel the *same thing more severely* to night than at any time during the day. Therefore your soul-principle is not very much disturbed, is not struggling vehemently to expel the invader, except at that particular period of the evening or night. But at the appointed hour, prompt and true as the clock on the mantle, your soul-energies summon all their available forces to overthrow the enemy. The Self-Healing principles exert their wisdom and their power alike for the reproduction of harmonial conditions. But suppose you—the yet more interior Man—neglect to "second the motion." What then? Suppose you, either thoughtlessly or culpably, withhold your encouragement. Suppose you do not exert your Will to cheer on the struggling soul in its gracious efforts to abolish the "cold," what will be the consequence? ANSWER: An increase of the original disturbance, a deeper penetration of the enemy, a more hampered condition of the physiological functions, a more complete overthrow of the soul energies.

This is the biography of every case of chronic disease. Consumption, dyspepsia, &c., are generated in no other way—namely, by a disregard of the *periods* when the Self-Healing Energies (the soul-principle) exert themselves for the re-establishment of right physiological conditions. Look out in the fields. The earth performs her functions *periodically*. She joyfully receives the prolifications of the sun every spring; her beautiful bosom enlarges from month to month; the rivers of life flow through all her veins and arteries; she becomes pregnant with a million harvests; they are born amid thunder, and lightning, and storms; she then throws her autumnal and winter garments upon her offspring, clasps her myriad darlings yet closer to her bosom; then she feeds them lovingly till spring returns with all its impassioned smiles, when she again responds to the engerminating visitation of the sun and the showers. All is periodical. Night and day, seed time and harvest, disease and health, reciprocating each the other's life. Disease is but the disturbance of the soul-principle. But the effort at equilibrium is made squarely to the law of periodical changes. And we hold that all true treatment for the healing of all diseased organs or parts will be governed by the *kind*, *period* and *degree* of effort which the soul is putting forth in the physical economy. When you suffer most, the Soul is then making the strongest effort in the direction of Health. Of course, therefore, your soul-principle is struggling less with the disease whenever the disturbance is least realized. The middle state is a sort of temporary health-period. During such a period never take either medicine or magnetism. Be very still, or else resume your occupation, but always do everything with reference to *the period of struggle.*

CHAPTER III.
THE MEDICAL VALUE OF CLAIRVOYANCE.

The judicious employment of Clairvoyance, in the diagnostication and treatment of Disease, is a legitimate use of the power. The best natural judgment, though crowned with the diplomatic glory of a scientific education, is often incapable of reaching with certainty below physical disturbances to their primal causes. In the detection of the hidden sources of human misery, and of the conditions that generate corporeal discords, no sight less penetrative than that of the genuine clairvoyant can ever avail much. No professional man is more willing to acknowledge and publish this incapacity among medical men, than the vigorous editor of the *Scalpel.* We are ever and anon refreshed with his Carlylish style of proving and showing up the emptiness of the pretensions of the regular Faculty. The impossibility of discerning the deep-seated causes of many diseases, by mere sensuous observation of symptoms, is very generally confessed. The ordinary inferences, drawn from indications external, are frequently erroneous. The world of sick and suffering people can attest to the truth of this assertion. They groan day and night, or hobble about with congenital deformity, because their parents were once *patients.*

And yet the careful instructions of the scientifically-trained judgment are to be preferred as superior, and as being more in harmony with rational sense, than the blunderings of undeveloped or non-medical clairvoyants. There are many and various kinds of seers and seeresses. Only the few, however, of a certain kind, can truly diagnosticate and divulge the causes of Disease. The real sources and philosophy of human suffering are discoverable only by such of the seers as possess an appropriate faculty. The condition of seership is one of the greatest impressibility. It is too apt to take on and reflect the fears, surmises, or established convictions, of the patient. Every sufferer, whether blessed with intelligence or not, will entertain some *definite* conclusions regarding the nature and probable *cause* of his misfortunes and diseases. And the clairvoyant is very certain to become involved therewith, and will be misled by contact with the dominant feelings and judgment of such a patient, unless, as above mentioned, the seer or seeress be in the full self-possession of the faculty of sight, while in the act of diagnostication.

It is unphilosophical to suppose that all clairvoyants are equally or similarly endowed. It is rarely the good fortune of any one genuine medical-clairvoyant to possess abilities commensurate therewith in other departments of investigation. In fact, a first rate seer of disease is seldom more than a second or third rate prescriber. This incapacity is sometimes manifested immediately subsequent to an examination and prognosis which have been pronounced satisfactory. On the other hand, it sometimes happens that only the faculty to *prescribe* is perfect. That is to say, certain seers and clairvoyants can survey the field of nature and detect the exact *remedy* for a disease, the origin, location, or symptoms of which they professed no power to discern. In such case it is wisdom to obtain your diagnosis from one seer and your remedy from another. But this course presupposes your belief in the existence of the power. With this faith you can avail yourself of much benefit by candidly informing the *Prescribing seer* what the *Describing seer* reported your disease to be—the remedy for which is what you now seek.

Certain faculties are peculiar to certain temperaments. For example, some persons are only perfect in the *sympathetic* detection of your disease; while others, though obtuse as stone jugs to the lines of sympathy, have brilliant powers and fascinating gifts in other directions. But here let it be observed, that the *degree* of the endowments, as well as the kind or class, must enter into your account. A high development of excursional clairvoyance is incompatible with a low employment of the faculty. If a mind is endowed with the power to discern objects through the mystic distance—if it can discriminate between fancies and facts—then it is a *misuse*, a sad devotion of that mind, when its powers are pressed into the treatment of bodily diseases. Of course the reverse is equally true; that is, to urge a purely medical seer to probe Space in quest of distant objects of interest. The penalties of mis-employment, in either case, are mistakes and doubts.

In the flowings of inner life, it may happen that the seer accomplishes a silent spiritual unfoldment. Its immortal attributes may one by one bloom in the garden of the soul. In its clairvoyant proclivities and exercises there may occur a sort of apotheosis; an ascension of the ordinary powers to rank and fraternize with sublimer uses. The whole mind, pulling its attachments to material things up by the roots, may take interest in things trans-mundane. The seer may thus silently and unconsciously advance to the perception of great questions, of truths, of principles, of ideas. What then? Why, then, the investigation of Disease is next to impossible. The divine law of seership on *that* plane is repealed, so to speak; and the clairvoyant is no longer useful on the corporeal side of humanity. Any such use of the powers of a seer thus unfolded, would be attended with penalties not less severe than visual degradation. The transgression is against the operations and requirements of a divine law. In many cases mere worldly wants, necessities, or poverty of the moral faculties, have urged very high seers into the most unprofitable forms of the medical business; or still more external, into telling fortunes, reading the stars, psychometry, &c., to the exclusion of those excursional and spiritual exercises which expand the soul and develop its latent abilities. Such transgressions are attended with a loss of virtue in the spirit, a retardation of the normal processes of growth, and sometimes they ultimate in a total suspension of the clairvoyant faculty.

But we began this Chapter with the impression that something must be said on the subject of scientific errors and medical fallibilities;" and that we must pledge ourselves to render many installments of such service for the sake of mankind. In the department of physical suffering we have had ample experience, sufficient to make us "feel another's woe," and the lesson is not lost. In the diagnostication of disease, too—and in the practice of pilgrimizing through creation's empire in quest of remedies to remove disease—our experience is not less manifold and available. Nor are we this day negligent of "golden opportunities" whereby fresher inspirations flow through the temple of reason. We do not, however, employ the power of clairvoyance for the benefit of individual applicants. The reason of this was suggested in preceding sentences—viz: because such use and application would check our growth and usefulness in more important directions. Not only this, but because there is imminent danger in mis-employing a high faculty. A power of comprehensive vision may be impaired by mental unfaithfulness, or devotion to small things, on the same principle that shortness of natural sight is frequently caused by too constantly looking at near objects, and at fine points, by which the scope of vision is permanently circumscribed.

CHAPTER IV.
NO INFALLIBLE REMEDIES.

THE enlightened mind need not be told that an "infallible remedy" for every physical disturbance is an impossibility. The vaunted potency of certain empirical compounds, each prescribed and puffed by its particular inventor as a "sure cure" for every disorder, has well nigh disgusted the reasonable side of mankind. That there are virtues in patent vices no one doubts. Pills and plasters do some good in particular instances, and from such they derive extensive notoriety and popularity, but the day soon dawns when they fall into disuse and contempt. The fact that every invention embodies some curative virtue, and the additional fact that every system or school of medicine is as many times defeated as it is victorious, has led to the development of unbounded skepticism, and also to the belief in what is termed "Eclecticism."

The Eclectic school of physicians are both dissenters and receivers on a boundless scale. They avail themselves of every long-tested and established remedy, irrespective of the system in which it was originated, and prescribe it in all cases where experience proves its applicability. But the Eclectics do not master disease. They reject the evils of time-honored theories, they look upon human suffering with clearer eyes, they classify diseases and assort remedies with all the medical experimentation of twenty-two centuries to simplify and exalt their labors, yet it must be acknowledged that this new school has neither achieved any very marked triumphs nor sustained any very severe reverses, but in general is *as successful* in the treatment of disease as any of the recently developed *myths* and *pathies*, and that is all we can now say.

Nevertheless, we glory in the development of "Eclecticism." It implies a vast scientific independence; a breaking up of the old mineral consolidations in medicine. It says: "Investigate the laws of matter in man, rise above the historical horizon of physiological knowledge, think for yourself, yet read the works of Hahnemann, Dixon, Thompson, Bell, Muller, Hall, Hunter, Marshall, Bischoff, &c.; for in the fullness of human research and conclusions ye find the *golden mean* of common sense and medical prosperity."

But physicians of every school can cure more than they do. The difficulty is, that their patients will not follow the tenth part of dietetic hints and rules which intelligent medical men prescribe with, or as a *part* of, their treatment.

It is our happiness to believe and repose unbounded confidence in the *Self-Healing Energies* that impregnate every fiber and function of the organization. In them we find the whole *Materia Medica*—astringents, tonics, emollients, corrosives, stimulants, sedatives, narcotics, refrigerants, anti-spasmodics, antiseptics, sialagogues, expectorants, emetics, cathartics, diuretics, diaphoretics, emmenagogues, abortives, antacids, errhinds, and the fifty other effects or symptoms which these awful names are scientifically designed to classify—all, yea all, may be found in that wonderful repository of health and disease, *the constitution of Man!* And we hesitate not to affirm, in justice to the hundreds of capable and honorable men in almost every school of Medicine, that if patients would, in sickness, conform strictly to *rules of health*, the doctors could exert far more mastery over the domain of Disease than they now do. The mass of physicians confess that the human system demands almost nothing besides good treatment, nursing, bathing, dieting, magnetizing, &c., during the different stages of any known disease. But the doctor is oftentimes compelled to sacrifice his judgment to the ignorant demands of his impatient subject. One prominent reason why the world is swarming with *chronic curses*, is the fact that, while convalescing from the effects of some severe attack, either in organ or function, the patient stealthily omits the helping remedy, and at the same time commits the private crime of over-eating and indulging in forbidden luxuries. The consequence is, that such patients rise from their bed of suffering seemingly restored, but really with the seeds of chronic disorder scattered all through their frames.

The effect of all this is self-evident. The imperfections and glaring failures of various medical systems daily strengthen the spinal column of a *skepticism* in regard to all medicine on the one hand, and develop a very general *Eclecticism*, or free selection, on the other; between which, *diseases* and *nostrums* will multiply and disturb the world, until the great mass of mankind, in the circle of progressive civilization, will "throw physic to the dogs," and thereupon commence the rational era of inherent healthfulness, based upon obedience to the *Self-Healing Energies*, which never fail.

If, however, any person should "flatter himself" that he can violate the conditions of Health, and, at the same time, by simply yielding to the self-restoring mercies of his spiritual constitution, recover all his original vigor and bloom, his disappointment will be complete. Mother Nature is as loving and as just as Father God. They do not, because they cannot, guarantee impunity from the effects of violation. All the medi-

NO INFALLIBLE REMEDIES. 33

cal *isms*, and *myths*, and *pathies*, from Hippocrates to the last nostrum, cannot perform the pardoning act. *There is no infallible remedy.* Pass the word all around the world—*there is no specific for any human transgression.* Let every ear hear it. Let every eye read it, and inscribe it in fadeless characters upon the temple of personal health; and let all the world rejoice that there is no safety in habitual disobedience. The best medicines are scarcely more than hints to Nature in Man. They oftentimes operate like oil on the wheels of life; but neither the broken wheels, nor the life processes impaired, can be restored by medicine. Let parents instruct children to comprehend this important truth—*no vicarious atonement*, but that every physical as well as moral transgression is visited, in due time, with a just and certain punishment. No priest, no physician, no medicine, no religion, no future, can alter the fixedness, or weaken the justice of this eternal principle. Therefore we affirm that all empiricism, either in "Law, Physic, or Divinity," is a mischievous deception. An infallible remedy for a physical sinner is just as impossible and deceptive as a vicarious atonement for a spiritual sinner. The only infallible way is the way of wisdom—straightforward obedience to the laws of Nature.

Although clairvoyance can detect the healing properties of minerals and plants, as it were by *instinct*, yet, wonderful and far-reaching as is this detective power, nothing is more certain than its occasional failure. Spirits may take a benevolent interest in the conditions of the sick and suffering, but they, too, many times, fail in effecting perfect cures. One spirit-seer, for example, prescribes magnetism in a particular case. The application is immediately and marvellously effectual, and the patient walks forth strong, healthy and prosperous; but on the morrow the same spirit prescribes the *same* remedy for the *same* disease in the body of another person, and lo! the failure is complete. Why is this? Because there is no power in heaven or on earth, that can, with unerring precision and unfailing judgment, hit the mark every time exactly in one place. Hence the reasonableness of our conclusion is made manifest, that no safety can be found outside of perpetual obedience to the established laws of life and health.

Palliations, however, are natural. If we cannot restore the limb, it may nevertheless be bound up and protected. The rheumatic joint may be unsocketed by long suffering, or in another case the morbid action in the different tissues may prognosticate ultimate decomposition, yet we do gladly behold possible aids for the one, and checks to the downward progress of the other. Celebrated investigators, such as Chomel, Bright, Andral and Louis, have informed the profession upon the uncertainty of any treatment in tissual diseases, except that of special watchfulness and nursings. We allude to "nursing," because it is an approach to the magnetic method, which should be studied and employed by every reform family. But let us now turn our attention to the operations of disease and debility in the human body, and let us walk in healthful paths.

We have knowledge of cases where Gen. De Bility is outrageously invasive and triumphant, notwithstanding persistent medical efforts to accomplish his overthrow and expatriation; but, on the other hand, we have authentic news from quarters where the unprincipled Fillibuster has been gloriously assaulted and vanquished.

Let no human being despair of ultimate victory. There is a beautiful sanctity in the temple of Health. Fear not, O Man! to leap into the rivers of happy physical life. Bathe thy spirit's material covering in the Euphrates of sweet flowing waters. The isles of gladness are fertile in the lake of harmonious existence, whose margins are blended with the bright blue of infinitude, and the waves whereof unceasingly roll with anthemal melody eternal. Fear not, unhappy one, to mingle with Nature's health-giving elements. Let the strife come at once! The battle of life is not fearful, is not dangerous and uncertain; for even now an angel's finger is raised to point the way to youthfulness and victory. No sentimental repining, no over-weeping, no under-creeping, no diurnal thanklessness; but up at once, in the stillness of the voiceless night, (if chased by ill-omened dreams) up! and lay no more down, until a more peaceful silence covereth all thy being. If a fever is upon thee, or if there be a heaviness upon thy breast, rise from thy bed of suffering, disrobe thy body, walk within thy room, where heaven's purest air should gently circulate from window to window, and, in the holy silence of the solitary hour, dip thy hands in refreshing water, and thy feet also, and manipulate thy whole person, with steady purpose to remove the heat, so that tranquil sleep may lay silently down upon all thy bodily organs, even as the divine blessings of immutable truth hover about the inmost soul. A body full of virtuous health—of harmony deserved—is a form of holy beauty. Remember the Poet's Scriptures. In them 'tis said that

> "A thing of beauty is a joy forever;
> Its loveliness increasing, it will never
> Pass into nothingness, but still will keep
> A bower quiet for us, and a sleep
> Full of sweet dreams, and health, and quiet breathing."

A few persons write us to say that, on attempting to exert their Will, through the pneumogastric and voluntary nerves, they become mentally very chaotic, unable to fix their thoughts upon any one locality, and a distressing weariness soon creeps upon the entire body. Other persons, with naturally more concen tration of mind and purpose, write us of the efficiency and efficacy of the Will. We shall now speak a few words to the former chaotic class. They are the most difficult patients to treat with magnetic forces; for, owing to the incessant jarring and vibrations of their nervous systems, they *fling off all influence* as rapidly as the strongest magnetist can supply and fix the power upon them. Dear sick reader! allow a few words to be spoken regarding the *routine* of thy daily life.

You are invaded by Gen. Debility. We suppose you to reside in America, where the climate is a hydra-headed temperature, driven by the finest spring and summer in the known world. We have on this continent a *mixed* atmosphere, a *mixed* temperature, a *mixture* of all the seasons in every section of the year; and we are, in consequence, a *mixed* population, with *mixed* religious opinions, *mixed* political systems, and a perfect *admixture* of health and disease. Under these recurrent changeabilities, it is more than simply unreasonable—it is pure folly—to expect a uniform condition of either the mental or bodily constitution. Both involuntarily sympathize with the conditions and circumstances of the world in which they for the present exist together. Colds, catarrhs, croups, and consumptions, are, therefore, just as natural to man—under pervading conditions of the globe—as winds, frosts, snows, tempests, floods, and droughts, were natural to the juvenile stages of the earth. But—mark you, one and all: *Man is a wondrous magazine of Self-Healing Energies.* So is the earth on which he roams. True, reader, but Man is a *thinking* and intelligent organization. The good earth is not. Man grows and unfolds from *within*, and consequently may draw upon the fountain again, unless by vices and transgressions he has drawn away too much of the restorative principle. Ah! this is the sad condition of thousands. Is there, then, no "balm in Gilead"? Is there no hope for such? No chance to replenish their exhausted forces? no "straight and narrow way" for chronic *sinners* to be saved?

We answer affirmatively, and bid all the diseased to enter upon the duties of the life-battle, by which Gen. DeBility can be at once and forever vanquished. And we would, therefore, admonish every sick one to read what we have already communicated, and to adopt whatever portion of the remedies is best fitted to individual necessities. But we have much to write respecting many functional diseases, not yet named in these chapters.

When the warm season is upon us, during which the early morning air is most magnetic and renovating, there is one straightforward habit to be acquired by the consumptive and digestively debilitated—that is, to rise in the early dawn, (say at half-past four, until autumn,) dress comfortably, and immediately proceed upon a ramble—*eating a small orange while walking*—and be sure to return, disrobe, and to lie down to rest one full hour before breakfast. Sleep for thirty minutes, if possible—it will be possible by practice—and rise in time to take a hand-bath, and to dress, all before the bell rings you to the first meal.

We know *how cruel* is this almost sovereign remedy for vital debility. It would be far easier to "give up"—to keep in bed while the golden sun is rising "with healing in its wings"—and, as the foolish neighbor does, to send for a doctor to administer doses of indescribable drugs and tinctures. But let your reason act, and it will inform you, notwithstanding the hardship of early rising, that a *walk*, a *sleep*, and a *hand-bath*, (wash your whole body as you would your face, in two quarts of cold water,) are the most glorious remedies for vital debility possible to prescribe The refresh-

4

ing SLEEP before breakfast, and after a walk in the morning, is, particularly in this climate, a draught of pure health. It is well, also, to sleep about thirty minutes before dinner. Let this rest be inaugurated by the pneumogastric effort and vital equalization. If, after you rise early, you should attempt to read a paper or book, to study, or to indulge intellectual exercises, the debility of your system, nerves, bowels, and blood, will be in a few days increased four-fold. Avoid, therefore, all *mental* occupation previous to eating the morning meal. Otherwise, your strength will soon be weakness compounded.

In regard to eating: It is proper to caution all debilitated persons, young or adult, in the summer months particularly, not to eat *fruit* and *vegetables* at the same meal. Together, in the stomach and bowels, they chemically operate to generate various semi-poisonous gases; by which dysentery and various bowel disorders may be produced in a few hours. Of vegetables, beside bread, use but two kinds at one meal. Away with the "stalled ox"—forbid the "fatted calf"—be superior to "fowl play" as nourishment—for your sake no sheep need be "led like a lamb to the slaughter"—let no swine "find favor in your sight"—but, without pastries and sweet things, live rationally and thankfully every day.

We admonish you to be perpetually conscious of the glorious world in which you live; but which, with all its beauty and glory, is but the alphabet of the Spirit Land in the surrounding immensity. By coming out of the Egypt of Disease, your eyes will open upon the universally distributed glories of the visible creation. Your happiness will be increased an hundred fold in a few weeks. Now, while sick and confined to hospital limits, you live like subterranean mortals, shut out from the glory of the infinite. Who ever knew a first-rate good *evil* man? Or who ever saw a happy and healthy *sick* man? Exodus and emancipation from Disease is a wonderful deliverance from the "Devil and all his imps." The temple of Nature would possess a new charm to the thus delivered. To illustrate what effect it would exert on the good-minded, we will quote, from Cicero's *Natura Deorum*, the following sublime conception:

"If there were beings who lived in the depths of the earth, in dwelling places adorned with paintings and statues, and everything enjoyed by those most wealthy and fortunate in the world; and if these beings could receive tidings of the glory and power of the divinities, and, after that, come out from their dark residences through the fissures of the globe to the surface on which we stand; if they could suddenly see the earth, and the sea, and the circle of Heaven—contemplate the great cloudy expanse, hear the winds of the firmament, and admire the majesty and beautiful effulgence of the sun; could they behold the starry host of heaven in the night, the rolling and changing moon, and the rising and setting of the celestial orbs in the order prescribed from eternity—they would surely exclaim: 'There are indeed Gods, and such magnificent things must be the work of their hands.'"

CHAPTER V.
SELF-HEALING ENERGIES BETTER THAN MEDICINES.

Earthly language cannot embody all we have to impart under this head, in regard to the perfect adaptation and competency of man's vital energies, to self-repair and harmonize the bodily organs.

As soul speaks to soul in the blissfulness and breathings of magnetic attraction, so the powers which live in all the cerebral centers and visceral organs meet and mingle together, like angels in the garden of light, for purposes of greatest good to the physical and mental proprietor.

Suffering Ones of earth! have ye not realized the medical wisdom that floats through every vein of your physical structure? There is no power more self-just and self-restoring than that which breathes, and sobs, and gushes, in your personal organization. Atheism is not more destitute of the divine qualities of intuitive wisdom than are the various systems of medicine of that restoring principle which alone can summon the spirit of health from its retreats in the corporeal economy. As ye cannot gather figs from thistles, so ye cannot obtain health from the drugs and medicines. Medicines cannot impart the principle of health, any more than can a book convey the light of wisdom. But there are aids and helps in medicine, just as there are hints and streams of suggestiveness in books.

The error, however, is in the source of reliance. The sin-sick soul goes meekly to a priest, or prayer-meeting, with unbounded confidence in the efficacy of the remedies and ceremonies prescribed by the priest. In like manner the diseased mortal seeks the doctor with unlimited faith in the power of popular nostrums and inorganic compounds. . And yet, as the age of honesty and intelligence expands, we find patients and physicians more and more agreeing that medicines, *at best*, but serve and subserve the inherent energies of the organism ; that health is possible only by means of the self-restoring and conservative principles which the good Father and Mother transmitted to the organs, muscles, nerves, and blood, of the living temple ; and, therefore, that all belief, or pretension, that medicines hold and convey the life-giving energies of health and beauty to man's body, is nothing less than mischievous superstition or intentional imposition.

Disease, in very shortest phrase, is *discord*. The causes and effects of this one "discord" are various and innumerable. They differ in different persons, because of temperament and occupation; also, they differ in different seasons, because of *temperature* and potential electricities. Spring-discords of body are different, in the same persons and places, from those which prevail in autumn; so, also, for reasons above given, winter-disturbances differ widely from those of the summer time.

Outward manifestations of invisible disturbances are indications of the causes of discord, which preponderate either in temperament, occupation, or temperature. Something is unbalanced in the empire of concealed forces—either an *excess*, a *deficiency*, or a *misplacement* of parts—perhaps, as occasionally happens, all these causes of pain and suffering exist and operate in combination.

When the body is thus besieged with " discord," how can the

soul feel harmonious? It cannot; for mind must suffer with the organs by which it exists. This fact, however, is of the highest significance. It teaches that the soul—which is the Fountain of forces out of which the mind rises into entity from an elemental state—contains the conquering and health-giving powers. From these energies, and not from medicines, the sick may expect relief. Sweet and grateful breathings from invisible principles are cognizable only by means of the sensitive energies of mind. Granite rocks do not hear the whisperings of infinite wisdom, although they invariably move in harmony and keep step in progress with such wisdom; while human beings, when unfolded in their affections for the soft and thrilling music of truth, not only obey divine wisdom in the simple luxuriousness of spontaneous freedom, but, in addition to such obedience, they receive and enjoy that handsome healthfulness which naturally results from equilibrium with the principles of such wisdom. It is true that medicines and magnetisms can and do contribute (sometimes) to the right development of the patient's inherent energies of health; and so, indeed, do all arbitrary appliances affect, either for good or for evil, all bodies and forces with which they come in contact; but this surely cannot be interpreted and appropriated as a sufficient *foundation* for the absurd doctrine that drugs, medicines, and magnetisms, contain, *per se*, virtues which may displace or regenerate your vices and consequent misery.

Nay, nay—listen, O suffering soul of earth! and thine ear shall catch the soft and grateful music of Nature's truths. The authority and divinity of thy Father are impressed upon each speaking principle. Motion, Life, Sensation, Intelligence—do not these principles rule, or ought they not to govern, in every part of your physical body? Do they not labor day and night, both when you wake and when you sleep, for the full growth, refinement, and harmonious expansion, of all organs and functions? Is not the unspeakable goodness of our eternal Father and Mother displayed within the temple of your individuality? What principle of wisdom was it that, before you had a personal consciousness, built your bones and jointed them so beautifully? Do you not suppose that the same energy is with you this very moment? Behold with what promptness your windpipe expels a grain of sand or a bit of bread, which perchance invaded the province formed only to attract and enjoy the pure air of heaven. With what divine energy does the spirit of blood work to heal the wound on the surface of your temple? A mote is in thine eye—instantly the tissues send forth a flood of tears to wash it away! Thy stomach is stronger in righteousness than thy brain. Ever since your birth the digestive receptacle has been tempted, fed—crammed, jammed and poisoned—with every imaginable good and evil thing, from the outrageous inventions of the nurse, who first took you from the doctor's hands, to the last pastry cook's indescribable compound of table temptations. Your brain could not contain one half the shadows which such substances cast upon it, neither could your judgment dispose of a tenth part of such feeling in the shape of thought and intelligence; but your stomach pours forth its fluid, and commissions its self-healing attributes so promptly and perfectly, that all you can justly complain of is—"indigestion;" headache, of course, and intestinal disturbances.

Recuperation is natural to all living bodies. Every derangement, if curable at all, the self-repairing energies may overcome with harmony. Only open your understanding to a knowledge of those magnetic agencies, or perhaps mechanical aids, which will remove obstructions, and thus put into Nature's hands the reins of physical government. Man's fearfully constructed body is more than paralleled by his wonderful mind. This power is endowed with the tendency to work both ways; it will either bring disease upon the body or remove it. If your own mind cannot *begin* the work, (which many times happens,) then avail yourself of the healing attributes of *another* mind. If your entrammeled imagination still needs a visible medicine to fix its faith upon, why—we are sorry for you—send for the only physician in whom you believe; and then straightway "let us pray" that he (or she) will give you the most artistically *small* dose of medicine possible for his sense of justice and adaptation to suggest.

The truth is, that, accidents excepted, *the great majority of human bodily diseases are of mental origin.* Disturbances *begin* in the *forces* and *end* in the *forms;* therefore, by virtue of a psychological and physiological necessity, the remedy must commence in the form and terminate in the spiritual constitution. Swallowing a disgusting mass of medicine is never necessary, any more than is a weekly dose of orthodox religion indispensable to good morals and happiness after death; and yet, disgraceful and disagreeable as it is, there are millions of our humanity who habitually take atrociously large doses of both! From all this, and innumerable other equally popular outrages, "good Lord deliver us!"

CHAPTER VI.
A PNEUMOGASTRICAL DISCOVERY.

MAN is an immortal, self-conscious spirit, enveloped for wise purposes by matter of a grosser sort, over which he is designed, through his volition and intelligence, to hold the supreme and exclusive control. His integral attribute for self-development and self-government surpasses the belief of his uneducated judgment. He is reckless and faithless in regard to the divine principles of his *inner life*, because he is boundlessly ignorant of the eternal riches that lie buried within the soil of his spiritual constitution, and the penalty is embodied in the form of DISEASE, whose miserable offspring are innumerable. Thousands of our brothers and sisters—good and tender-hearted people, who inhabit the hamlets and mansions of earth—are dragging out a wretched existence. The kindest, most skillful, best paid, and latest diplomatized physicians, can yield no perfect health to these "sin-sick" and suffering ones—and why? Because, in too many instances, both doctor and patient are equally ignorant of the *psychological energies* slumbering in the human organization. Coarse, cruel, crude, bitter *drugs!* What can they do for you, compared with the kindly offices of *the spiritual forces* that bivouac in every part of your wondrous organism?

Our new discovery is not to be patented. We design it for

the universal use and benefit of man. It is a self-adjusting and inexpensive invention, for the exclusive good of those who, from whatever parentage and subsequently producing causes, are summoned into the battle-field against the depredations and unlawful invasions of *General DeBility*.

The insidious operations of this celebrated "General" are delineated, and prescribed for, in almost every modern newspaper. "The Regular Faculty" proceed against him and all his embattled hosts, with countless systematic poisons, all established by Chemistry, and obscured from vulgar gaze by Latin habiliments—"too numerous to mention." Thompsonians strenuously insist that the old headstrong "General" dreads the formidable array of their forces—in classified battalions—*Lobelia, No. 6, Diaphoretics, Compositions, and Steam!* Hahnemannians, with indisputable argument, show that the most certain plan is to *fight down the first symptoms of insurrection*, invasion, or revolution—do this, promptly and scientifically, and the valiant "General" will depart with his armed host, overwhelmed and mortified with irreparable losses and defeat. The Cold-and-Warm-Water Brotherhood, with unfaltering zeal and undoubted intelligence, oppose themselves antagonistically to Gen. DeBility's encroachments with various streams and dispositions of the universal element—also, with napkins, compresses, bandages, straight-jackets, wet sheets, &c., to which the enemy *generally* expresses indignant messages. He grows daily more restive under such Guerrilla treatment, proudly disdains the style of warfare, frowns upon the hydraulic establishments throughout the continent, and retreats to more safe quarters and mountain fastnesses, viz: to the families where the "Regular Faculty" make their regular visits.

In these remarks we do not design to misrepresent any system of medicine. We affirm that no medical system is master of that renowned and universal conqueror, Gen. DeBility. Is this statement a misrepresentation? We have knowledge of skillful physicians, each with an extensive and lucrative practice, in whose families the General is an ever-present ghastly guest. We ask: "Doctor, why do you not prescribe a remedy for your diseased wife or daughter?" And he honestly replies: "I have tried every supposed specific, every known remedy for such cases, and each application or administration only serves to aggravate the symptoms."

Well, then, try the new medical discovery announced at the beginning of this chapter—not Kennedy's, recommended for every disease the human skin is heir to—but Nature's *pneumogastrical* remedies, with which, whether at home or abroad, you are always abundantly provided.

What do we mean? We will explain briefly as possible. Man's fearful and wonderful organization is regulated, or may be, to an almost unlimited extent, by two great positive and negative conductors, viz: The sympathetic and the pneumogastric nerves, which, like all the lesser and more delicate nerves, take their rise in the brain, which is the fountain-center of all vitality and sensibility.

Of the first conductor, the Sympathetic Nerve, we will not now write anything at length. It is a wondrous and beautiful magnetic telegraph cable, consisting of a connected series of polar centers, or *ganglia*, distributing branch offices through all the internal organs, communicating with nerves both spinal and cranial, and conferring, by transmission, the principles of life and sensibility to all parts of the organism. The reason why we do not speak of this great Nerve in this article is, because, according to our careful and oft-repeated examinations, it belongs to the great automatic hemisphere of mind, and will be the very last (because it is the highest and best) to ascend the throne of Will. Instinct, impulse, soul-life, heart-essence, "the depths of the soul," as will appear in our next chapter, are characteristic sensational attributes of the "Great Sympathetic Nerve." This position is sound—psychologically, physiologically and logically sound—therefore, we do not fear to write it, nor to submit it to the candid consideration of every progressive or regressive physician.

The Pneumogastric Nerve, on the contrary, communicates no essential life or vital sensibility. But it confers upon the internal organism vigorous energies and voluntary movements. It is in direct and intelligent contact with the attributes and designs of Will. It is, consequently, a voluntary agent of the cerebral intelligence, a motor nerve, so to say, by which the Volition may express its decrees upon the whole physical economy.

THE DISADVANTAGES OF IGNORANCE.

The more ignorant the mind, the cruder will be the manifestations of its desires and ordinances, and the more *in bondage* is such an one to the voluntary nerves; while with interior knowledge come the requisite powers and purposes to rise out of embarrassments into comparative freedom and independence. We make these remarks because, as we too well know, there are thousands of so-called intelligent persons, occupying high official stations in every kingdom, whose physiological experience is in direct antagonism with what we affirm respecting the nature and capacities of this Pneumogastric magnetic conductor of the voluntary energies. But if such, together with all doubting readers, will perseveringly experiment on themselves, while desiring truth and health, we know that we shall receive their unqualified indorsement.

Become clairvoyant a few moments, suffering reader—take

the living subject, (your friend,) look straight into the constitution of his voluntary system—and the facts, with regard to the origin and function of this Nerve, will shine out upon your reason. Keep your vision steadily upon the nether portions of the brain. See! a multitude of filaments put forth from the side of the *medulla oblongata,* which, as a primordial nerve, is a million times more sensitive than the brain itself. These thread-like filatories immediately blend into two small, white cords, filled with cerebral substances (which appear as *one* conductor) forming a ganglionic battery directly beneath the cranium, and thence descending through the cellular envelopment of the carotid artery into the neck, and downward, between the subclavian artery and vein, into the pulmonic organism, and terminating in various branchlets of nerves in the digestive system.

See, too! as this mysteriously philanthropic conductor of thought, and will, and emotion, proceeds into the far depths of its mission, how it lovingly twines around neighboring veins and arteries, and how benignantly it divides its incalculable wealth into private *fortunes* for the voluntary parts of the throat, heart, lungs, etc. The big names for all these benevolent bestowments and bequeathments of the Pneumogastric Nerve, are: *pharyngeal, laryngeal, cardiac plexus, and plexuses pulmonary and œsophageal.* Thus, in plainer language, the Pneumogastric Nerve, immediately on leaving the cranium, makes its will, and sets apart private fortunes (in the shape of voluntary powers,) for the benefit of the mouth, the speaking organs, the swallow, the heart, the lungs, finally pervading the entire stomach.

Now, let it be remembered, these are the parts or organs most diseased among men. These are the fighting grounds chosen by that popular Commander-in-chief (of very many

American women) Gen. DeBility. He marshals the minions of disease from their hiding places, and parades them in battle array upon just those territories in man where the pneumogastric conductor is wisely and wonderfully distributed. Consequently, if mankind but knew it, each mind holds the power, the long end of the health-lever, in the hands of his own Will. At a moment's notice the mind may summon its military battalions. When these powerful forces concentrate and commence action, under the command of an enlightened Reason, the myriad hosts of Gen. DeBility will at once speed away, like affrighted chamois at the hunter's approach.

Do you doubt, suffering and frail reader? Then doubt no longer, for see with your own vision the reasonableness of all this, which is in store for all.

We shall show you that bile—for the most part—is the broken down globules of the blood, and should be expelled. Food contains carbon and hydrogen. Old and waste atoms of the body are particularly loaded with the deadening and life less carbon. This is the ammunition—the powder and shot—used by Gen. DeBility. He takes deadly aim at your every organ, and fires away! His shots take effect, in the first instance, upon the *weakest* parts, either internally or externally—whether functional or constitutional—and down you go for two or three days, (or as many weeks perchance) unfit for the duties and manifestations of a Man! Now, although we still recommend anti-bilious remedies to aid Nature, we ask your Will-powers to exert themselves righteously, and the following is the initiative:

On the established fact that the Pneumogastric Nerve is a magneto-motive conductor of mental decrees, we commend you to yourself in the treatment and CURE—yes, *perfect cure!*—of almost all throat, lung, heart, and bowel diseases known on the American continent. But all cures are conditional.

The true causes of physical heat are not known. No chemist can perfectly explain the hidden sources. The incidental phenomena of heat and temperature are well enough comprehended, but the prime sources thereof are enshrouded in mystery. Fire is not essential to the heat and vigor of blood. The human blood maintains a temperature of about 98 degrees in all parts of the globe; the same beneath the equatorial sun as at the extreme north, amid eternal snows; and who, without our philosophy of the human constitution, can explain *the cause* of this amazing equilibrium of temperature?

We declare that it is within man's will-power to manufacture and preserve his temperature in any climate; and more, that he can maintain his health under any reasonable combination of circumstances; notwithstanding the many trials, and privations, and labors to which he is involuntarily subjected.

The common theory is, that by respiration the system receives and gives off the elements of heat. We say that this is the *process* of generating temporary heat, but the cause thereof lies in the spiritual principles composing the soul. (Let us pass on, leaving this subject for the present.) Oxygen of the air, entering the lungs by inspiration, mingling with the carbon and hydrogen of the blood, is supposed to disengage and distribute heat through the system. This is not the pure heat of love— the eternal *warmth* of the heart—but it is that which is wholly calorical and fleeting.

Instead, therefore, of breathing or eating for the purpose of increasing the amount of "animal heat," we would counsel *respiration* as a means of transmitting spiritual vigor to the weak and debilitated organs. The heart is covered with the *cardiac plexus*, which arises from the Pneumogastric Nerve. The lungs are supplied with many branchlets—another plexus of nerves—which, also, spring out of the same prime conductor.

Lastly, the entire digestive functions are pervaded and provided abundantly, and in like manner, from the same voluntary battery at the base of the brain—*just under the* WILL! "A hint to the wise is sufficient." The organs and parts named are under the immediate control of the enlightened mind. This is our "medical discovery"—try it, Brother and Sister—and keep trying, until you rise *redeemed and sanctified* by the blessing and grace of the God within you.

But how commence? First, if your weakness be general, and the blood is loaded with cold matter, lay flat down on your back, and, while breathing *deep, and slow, and uniformly,* WILL YOURSELF TO BECOME HEALTHY—in your feet and hands, in your knees and elbows, in your hips and shoulders, in your bowels and liver, in your lungs and brain! The heart will take care of itself. In cases where the weakness is generally distributed, all you are required to practice (while so prostrated and respiring) is the art of concentrating your Will and desires simultaneously on the extremities first; then work upward and *inward* progressively; and when, in the lapse of ten minutes of steady, deep breathing, you have reached the brain, repeat the process in the ascending scale, as indicated in the manner aforementioned.

By this Pneumogastric treatment of yourself, you will receive *spiritual* strength from the air—nothing is more certain! When, by practice, you can *breathe* deeply and heroically, and at the same time put your Will upon the restoration of the general system, the art of fixing your mind upon some particularly diseased part will become less and less difficult. *Consumptive persons*, by simply breathing profoundly, and willing systematically, *may enlarge their chests and lungs beyond the possibilities of disease.* Persons of cold temperature, with irregular habits and bad practices, may "right about face" and become harmonially healthy. Learn to depend upon yourself—use the infallible remedies of Nature—and, in spite of priest or doctor, you will "pass from death unto life."

TIME OF EXERCISES.—In acquiring this psychological power over the destinies of your bodily state, and in becoming a Self healing Institution, whether home or abroad—it may be necessary to practice (either while on your back, or standing, or walking, or riding,) perhaps three times in each twenty-four hours. Never just before meals, nor soon subsequent to them; but the true time is when chylification is going on; about 90 or 120 minutes after eating. The spirit world will aid you, by forming a secret conjunction with the *pneumogastric* conductor. It is certain, gentle sufferer; do not permit yourself to doubt. Nothing is too good in Nature, in matter, in spirit, or in truth.

ORDER AND SYMPATHY.—It may, however, be indispensably necessary, in some cases, to avail yourself, at first, of the sympathy and daily order of some loving associate. Or, in some instances, it may also be necessary to secure a residence at some Electro magnetic Institute, or Water-cure, where the food and fluids for patients are prepared more in accordance with the rules of health and life. But it is far better to *convert every home into a Health Establishment.*

Arouse, my countrymen! Shake off this contemptible *incubus* of fashionable life, by which the good of our human nature is often changed to bitterness. Abolish all desserts from your tables. Never eat more than three kinds of solid food for dinner. No drinking while eating. Masticate slower. Drive all complainings out of your homes. Do good to all; harm to none. The best blessing to ask "over meat" is, for a cheerful and contented spirit with which to digest what you swallow. All solemnity during meals is as irreligious as excessive mirthfulness is vulgar. In short, suffering children of earth! let each so live—

> "that when the summons comes to join
> The innumerable caravan, that moves
> To that mysterious realm, where each shall take
> His chamber in the silent halls of death,
> Thou go, not like a quarry slave at night,
> Scourged to his dungeon; but *sustained and soothed*
> *By an unfaltering trust, approach the grave,*
> *Like one who wraps the drapery of his couch*
> *About him, and lies down to pleasant dreams!*"

CHAPTER VII.
PATHOLOGICAL OFFICES OF THE SYMPATHETIC GANGLIA.

It is deemed best, since we are not writing these chapters for the exclusive benefit of educated physicians, to express our thoughts in language suitable to the comprehension of the common mind. Now and then, however, we may be permitted to employ technical terms, because they best embody the thought designed for the reader's understanding. And here we will add that we hold ourselves open to "more light," to the end that we may explain and further illustrate what has been freely, but perhaps too vaguely, communicated.

All the knowledge in our possession concerning the existence, anatomy, and pathological offices of the Pneumogastrical Nerve, was acquired by means of clairvoyant examinations of the human body, extending through a series of years, and under every imaginable degree and variety of mental condition and external circumstances. Much, therefore, is a matter of memory, from which we draw perpetual lessons, as from the well-spring of a strange and multifarious experience. But while writing upon these medical questions, as well as upon other topics, there is, in addition to this available treasury of past examinations, a present *illumination* of the intellectual faculties, and a finer clairvoyance also, whereby the things in human bodies, about which we write, are made limpid and as systematic as the trees in the landscape and the flowing waters thereof, under the beautiful effulgence of a July sun.

Of the Pneumogastric Nerve, we trace its magnetic terminations into the intestinal cavity, upon the muscular fabrics of which its motive influences are freely exerted. This fact last named, which is most perfectly exhibited to the inner vision, might induce the erroneous impression that the Nerve itself penetrates the lower abdomen, while, in fact, it reaches and centers only in certain portions of the upper viscera, but *practically*, (*i. e.*, in the effect of Will upon this Nerve,) we were perfectly correct in our last chapter, which the reader is requested to re-examine and analyze.

The internal beauty of the human physical temple surpasses the descriptive power of language. Viewed with the leaden eyes of materialism, and studied from the wholly physiological standpoint, the body presents nothing either "fearful or wonderful." Once open your spiritual eyes, however—fix their analytical powers upon the anatomical structure of the nervous system—and the *wonders of a universe* are instantly unfolded to your understanding. Wheels within organs, tissues within muscles, fibers within nerves, globes within blood, motion within life, sensation within motion, and myriads of beautiful processes going on in the several departments of the temple at the same moment—all impress the spirit-observer with sublime and unutterable truths, and with gratitude beyond all bounds of expression.

The principal mystery of the nervous system is its invisible influence or energy. Physiologists have long since supposed an identity between the so-called galvanic fluid and that potential principle which pervades and imparts divine dynamics to the whole nervous organism. But we are now enabled to approach the presence of more interior realities, and to comprehend more truth respecting the nervauric mystery of man. We are permitted to contemplate the human system in its perfectly *dual* structure, and to behold the labyrinthine passages and sequestered retreats of the Interior Life of Man. The object of all such discoveries is the harmonization of the individual, with special reference to the banishment of disease and the establishment of pure health, as the basis of all spiritual prosperity and progression.

The double structure of man's constitution is illustrated through all its wondrous detail. Every nerve, however thread-like and delicate, is composed of two distinct cords—positive and negative lines or conductors—each having a separate and distinct function to perform in the organic economy. The positive nerve is filled with a conducting substance essentially different from the material within the negative nerve, both lying side by side in one membranous sheath, and discharging different duties with the most perfect harmony and reciprocation. It is beautiful to observe how all the nerves start out from the primitive brain, (the *medulla oblongata,*) and proceed in the performance of their mission in married *pairs*, two by two, and how they never become confused and separated so long as body and soul remain together.

The encephalos and its nerves constitute the basis of all physiological existence. From the brain we proceed downwards and *inwards*. The spinal nerves, each brace of which connects with the spinal cord, are systematically arranged in pairs with reciprocal offices. One side of each nerve—or rather one nerve within each tube—conveys *motion;* the other, *sensation*. These motory and sensatory nerves start out from the brain-matter, and extending down the whole length of the spinal column, they branch out upon the several internal organs, and distribute themselves to all external parts of the system. Besides this general statement, with respect to all the nerves, it should be observed that there are two systems of nerves, which operate differently within the temple.

First, there are the long nerves and the plexuses; second, the sympathetic nerves and ganglionic centers. The first ramify throughout the exterior parts of the frame, and are the direct lines of sensational experience, connecting the intellectual part with the external world. Solids, fluids, sounds, odors, ethers, elements, objects—all get themselves reported upon the brain by means of the five senses, and the nervous influences by which the organal parts of the system are perpetually inspired. The ganglionic system, on the other hand, is an *internal* structure for the direct maintenance of organic existence. The grand source of internal sympathy is sometimes termed, "the great intercostal Nerve," which arises within the brain, or from a joint contribution of the fifth and sixth pair, and descends by the sides of the bodies of the vertebræ of the neck, thorax, loins, and os sacrum. In the thorax it perforates the diaphragm, forms the semi-lunar ganglionic plexus, and proceeds to ramify, and fix ganglia, in all the abdominal viscera. In fact, there is no part of the body neglected by this great *sympathetic* conductor of automatic principles.

This great nerve-system within man's body is the connecting link between lower life and the instinct of the spiritual constitution. It is quite different from the Pneumogastric Nerve, which conducts the will and decisions of the judgment to the heart, lungs, stomach, etc. The sympathetic Nerveplexuses, on the contrary, collect and convey the pure automatic principles of integral motion, life, and instinctive energy, to all the interior structures, to all the lymphatic vessels, and to every particle of living blood in the organism, by means of innumerable filaments; so that, in a certain and positive sense, all parts of the body are sympathetically related and tied by the bonds of affection together, forming one brotherhood of interest and mutuality of functions, and making it quite impossible for one member to suffer without disturbing the health and prosperity of all other parts.

We repeat that the Sympathetic nerve confers nothing in the form of sensibility or power of movement to the organism; and yet, what is far better, it is the grand conductor (almost *fountain*) of vitality and instinctive justice to the different and subordinate parts. Intuition is derived from the instinct of this wonderful system of ganglia. Sensibility, excitability, and irritability belong to the other system of conductors and the brain; but vitality, animation, instinct, and affection, belong to the great Sympathetic department; and so complete is the inter-mechanical operation of these nerves and ganglia, that intelligence and will are not necessary to the performance of their appropriate functions. The pathological offices of the sympathetic plexuses, however, are measurably within the jurisdiction of the possessor, as ultimately the whole interior will be subordinated to the voluntary powers of the cerebrum, when man will put all diseases, as well as more hurtful enemies, beneath his feet. The offices pathological are involuntary, yet the vital ends are accomplished better when the individual will second the operation.

It is well known that in the healthy brain there is a considerable quantity of phosphorus. In idiots this element is deficient. Phosphorus not only exists in the tissues and fibers of the brain, but this element is constantly and incessantly secreted throughout the entire ganglionic systems. The ganglionic globules are supplied with it, and it is in part by means of phosphorus that the blood is empowered to eliminate soul-aliment for the use of the brain and nerves. The direct and reflex action of the internal nerves, whether the mind is sleeping or waking, and several metamorphoses at the looping terminations of the involuntary conductors, are referable to the electro-chemical action of phosphorus in the brain and blood. Many physiologists have supposed that even the "nervous influence" is generated by the oxydation of the vesicular tissues, which is regarded as one of the four metamorphoses accomplished in the empire of ganglionic jurisdiction. The just and healthy action of the lymphatic system is inseparable from the vigilant operations of the cerebro-sympathetic nerves.

By means of these internal Nerves the cerebellum is apprised of any and every transaction in the vital department. Thus, for example, if any nauseous and poisonous substance be swallowed, the irritation occasioned by it in the stomach would instantaneously be reported to the brain, which, in its turn, would rouse the intellect and the involuntary system, and each is forthwith summoned to "fly to the rescue" of the afflicted part—while headache, dizziness, prostration of the will, and exhaustion, are effects very likely to ensue. In this manner one part telegraphs to all other parts through the great Sympathetic Nerve and the brain, which is the central agent and righteous ruler over the empire. And now let the remedial benefit of this be well understood What is that benefit? It is this:

The intuition of the Sympathetic Nerve is wiser than the best physician. Should water get into one of the air-passages, the Nerve says—"expel it!" and your judgment obeys the impulse to "cough it out." When your finger is bruised, the Nerve says—"bind up the injured member," and your judgment responds, "protect it," but perchance you disobey. The penalty in some cases has been "death by mortification." You are disposed to exercise when chilled, and admonished to sleep when fatigued, because the Nerve's intuition so dictates to your judgment. Do you obey? Or, instead, do you expose yourself to unhealthy temperatures, and stave off the natural sleepiness by artificial heat and ephemeral means of stimulation. If so, you do not harmonize with the intuition of the ganglionic system. You are guilty of disobedience. When diseased, the Sympathetic Nerve says—"Rest, rest, rest; be soothed by magnetism; let the lymphatic vessels operate upon and purify the adipose matter in your system: do not eat nutritious food now; be patient; let time work upon you." So the intuitive *ganglia* prescribe for the prostrate patient—not so the educated physicians of the land; though many of them know that this course would be far best in the majority of cases.

The pathological value of the Sympathetic Nerve is exhibited in its lessons of what is best and most needful when diseased. If the toad, the turtle, the mole, the bee, the dog, cat, horse, &c., (when left to themselves,) will properly prescribe the remedies adapted to their peculiar or accidental ailments, why not the more exalted and finely organized human being? Man's ganglionic system says: "Give me no drugs, but instead, gentle *aids* and *magnetic* principles." But that great experimentalist, the front brain, says—"Why not try a box of pills, a bleeding, and a blister?" And thus many times, when the whole Sympathetic system is crying out against the injustice, the voluntary experimentalizing brain decides to "try the nostrum." Once begun, it is hard to prophesy the result. So, then, since man cannot immediately affect the *ganglia* of the Sympathetic system by his will, let him at least permit his judgment to be instructed by the wise intuitions which are thus telegraphed to his sensorium and thinking faculties.

CHAPTER VIII.

PHYSIOLOGICAL VIRTUE.

Go to the superficial, and therefore *pompous*, teacher or *professor* of Physiology and Health, and he will merely educate your perceptive organs and train your memory. It is the inherent tendency of his school to educate and store your retentive faculties by means of isolated facts and multifarious observations. The experimental and *never certain* character of his outward science and skill, is, therefore, inevitable, and beyond the necessity of logical illustration. Even the most unlearned—the unpretending and common mind—can discern, at a glance, the *unreliableness* of much of the so termed medical science of the day. The fact, we believe, is conceded, that very few diplomatized and college-bred physicians pretend to *master* any of our continental diseases—such as Dyspepsia, Hepatic Disorders, Scrofula, Rheumatism, Erysipelas, and Consumption.

We trace the secret of this impotency, among medical men of learning and research, to one cause, namely: the Professors of our colleges of medicine—with few, but glorious exceptions —*take the student out of himself*, as though he were a spectator, a foreigner, a secondary and subordinate fact, to the science of health and the uses of medicine. The Regular Faculty seem shorn of the natural faculty of truth-seeing. They are scholastic in the department of ritualism, of formulas, of routine, of *fact*arianism; they can remember and quote illustrative remarks, from this, that, and the other medical authority; but how impoverished and used-up, how unscientific and shallow, how wordy and flatulent, when we call their attention to the deeper truths, to the sublimer realities, to the philosophical principles of Life.

On the other hand, to see the undisguisable contrast, observe how naturally the harmonial teacher of health approaches his subject and the student. He establishes, to begin with, the common profound principle, that "Health is Harmony"— that any, even the least, variation or departure from this fine balance and adjustment of the vital energies, is "Disease." If this departure and derangement be recent and severe, it is termed "Acute;" if of remote origin, congenital, or superinduced by violations upon healthy organs and conditions, it is termed "Chronic."

And now observe further, how the harmonial teacher of Physiology and Therapeutics appeals, not to the student's perceptives and memory merely, but to all the groups of organs, which are the physical foundations of the temple of Reason! He interests the pupil in himself; he attracts and brings him home. He then opens up to his intellect the realities of his own wondrous constitution. He explains the marvellous harmonies and fair proportions of the physical organization. He persuades the student out of his books into himself, and reveals the invisible fountains of recuperative energy, that rise and fall, that repel and attract, that expand and contract, that repose and labor, in the beautiful empire of individual physical existence.

What a glorious medical revelation! The great men of the Colleges pretend to *disdain* it, do they? Shallow pretense! Say, rather, that they *envy* the possessor of such simple, yet sublime knowledge of Nature and her Laws. Under the influence of this harmonial teaching, the student himself becomes a systematic work on Physiology. His self-healing energies constitute the most scientific Pharmacy of curative preparations. The inexhaustible treasuries of scientific lore are secreted in man's organization. The *true* physician, the unritual, but spiritual, teacher in the departments of physiology and health, is certain to reach the unseen springs of life, and he invariably depends upon the immutable flowings of vital energy for the success of his prescriptions. But patients seldom take any interest in the workings of the natural powers. Hence, frequently, to the efforts of the true physician the words of Pope are lamentably applicable :

> "Truths would you teach, or save a sinking land?
> All fear, none aid you, and few understand."

By physiological "VIRTUE" we design to imply, not a careless confidence in the never-failing operations of bodily functions—by which *abuse* and *neglect* are oftentimes surrounded and defended—but we mean the *co-operation* of habits and daily conduct with the requirements of the laws of life and happiness.

For example : If your lungs demand pure air, in order to circulate and purify the crimson current of life, you are *vicious*—morally and intellectually vicious—unless you supply that virtuous demand. If your mouth asks for bread and you give it *tobacco* instead, then you are not physiologically virtuous. If your body calls for rest, quiet, or a change of occupation, and you heed it not, or, instead, give it brandy and irritating stimulants, you are then violating the laws of organic virtue.

What follows ? All the vagabond troupe of *vicious feelings* which pervade, torment, betray, and crucify you, when you would be at peace within the temple. The reverse of these conditions is equally—yea, even more—impressive and forcible. That is to say, give your bodily organs the *free* use of whatsoever they *in health* demand, and cease feeding them while they yet have the still small power to cry, 'Hold, enough!" and your *virtuousness* will bloom beautifully out upon every look, word, and deed.

Rheumatism and analogous diseases are frequently caused by inattention to—*i. e.*, *vicious* treatment of—some of the expanding and contracting principles that regulate the organism. The two inseparable processes common to all animal bodies, termed *endosmosis* and *exosmosis*—first, the attraction of fluids and ethers from the external to the interior, and second, the repulsion of similar elements from the mucous membranes to the exterior surfaces—must be kept in a balanced condition, otherwise health is overthrown in an hour, and "disease" (of the sort natural to the person or the climate,) is the inevitable consequence. Vice, not virtue, prevails in such case. And the sufferer, like Job, is wicked enough to fancy that "the arrows of the Almighty" rankle in the marrow of his bones and in the nerves of his flesh.

A bad state of the liver is inseparable from evil impressions of men and things. A bitter-tongued and sour-stomached individual is no lover of music, though it may excite him, and his opinions of his fellow men will very nearly correspond to the state of his bodily vices. "The green-eyed monster" was never blest with fine digestion or a sweet breath. His food was changed into the "gall of bitterness," and his cup of milk into the tea of "wormwood," and equally perverted were all his impressions of men and the world. The doctrine of demons, of devils, of evil genii, was conceived in the womb of physiological vice. "Hell" is the shortest phrase to express "burning discord"—a great *boil*, on the way to suppuration—an inflammation of the brain, on the gallop to a hot delirium—an *erysipelas*, burning destruction into the flesh—an *itch*, with no power to scratch—a violent *discord*, resulting from physiological viciousness, is the bottom of the oriental conception of the "bottomless pit." The "pit" here referred to is no other than *the pit of the stomach*, whereon the great fulcrum of the lever of Health works, good or evil, just as the possessor, by his habits and conduct, may at any time determine. "Our young people," says a Thinker, "are diseased with the theological problems of original sin, origin of evil, predestination, and the like. . . . . These are the soul's mumps, and measles, and whooping-coughs, and those who have not caught them cannot describe their health or prescribe the cure."

CHAPTER IX.
BRAIN-LIFE AND LUNG-LIFE.

The mathematical and sympathetic correspondence between the visceral organs and different sections of the brain, briefly explained in foregoing pages, is beautiful and significant beyond the common understanding; and we know that we shall be conferring a permanent blessing by embodying, in a few more paragraphs, *the secrets and benefits* of these wonderful relations, processes, and representations.

Very intelligent and educated correspondents write us substantially as follows: "I have read your New Discovery for general debility, &c.; the principles laid down appear reasonable, but I cannot use my Will as you direct; my appetite is tolerably steady and good; my bowels operate regularly; I have no difficulty about sleeping; but somehow, *I am unable to gain strength;* my food does not build up any vital energy," &c., &c. Now let us consider the health gospel.

The tender spirit of many suffering ones blends to sadness and despondency. It is natural and righteous to desire to live long in the land. The harp-strings of the young heart tremble when disease seems determined to corrode them, to break them in twain, to pluck the dew-drops of happiness from the flowers of hope and health. An angel's visit is frequently interpreted to mean, "Earth-child, come, come away!" It is sad to yield up thy heart as a trophy, a slave, to premature Disease. Thou art designed and constituted for an earthly career " of three-score and ten years," and as many more as thy obedience to the Laws of Health may add thereto; but thou art not designed to dwell in the dark, damp dungeons of corruption and disease—not fitted for gloomy rooms of mortal suffering, while the earth is filled with splendors and music, and the heavens are trembling with the *soul-essence* of the Infinite and Eternal. There is nothing more desirable than pure, rosy, virtuous, meritorious HEALTH. O, that we could speak with the penetrative eloquence of an angel! Would that we could " dip our pen in the rose-light, fresh from fountains of the sun," and write out in a few comprehensive celestial sentences, the whole gospel of physical perfectibility as *the only basis* of spiritual completeness and endless prosperity! But we must content ourselves with the ink and language of earth, whereby to portray and enforce the glory and virtue of bodily health and greatness.

EFFECT OF AIR ON MIND.—Few persons imagine that their lungs are inseparable from their thoughts. Not that the pulmonary structures and functions occupy the heart of thoughts; but that as a man inspires the physical atmosphere, so does his mind conduct itself as to thinking, willing, and wishing. For example: If a human being should be imprisoned in a small room, not properly ventilated, and not replenished with fresh air from without—so that his breathing would be confined to the same atmosphere for a great number of hours each day—the consequence would unmistakably be exhibited in the mental operations of the victim. He would think in a circle, because he would breathe in a circle, and his digestion would be imperfect. His thoughts could not bound cheerily over the landscape, because the *atmosphere* of the landscape does not enter his lungs. Physicians and patients are habitually imagining that a "change of scene" is the secret of benefit in many cases of nervous prostration. Although there is truth in this impression, yet it is far from divulging the absolute paramount cause of the salutary results that sometimes follow pilgrimizing away from home in quest of health. When once the real *secret* is intelligently known, and when the knowledge accruing therefrom is promptly applied by the possessors, then may the multitudes of sick ones save themselves the fatigue and expense of journeys. If you wish to travel for recreation, first get a stock of health to sustain you, in the shape of Air, Light, and Electricity.

The shortest route to firm health is through the lungs and pneumogastric nerves. Small lungs—small minds; or, large lungs and bad air—large minds and few thoughts. The old-fashioned orthodox churches were built and kept as tight as drums during service; the effect was manifested in the narrow creeds and doleful doctrines concerning God and man. In this connection we are reminded of FLORENCE NIGHTINGALE, the noble nurse who voluntarily went to the Crimean war to bind up the bleeding wounds of the soldiers. She says:

"An extraordinary fallacy is the dread of night air. What air can we breathe at night but night air? The choice is between pure night air from without and foul night air from within. Most people prefer the latter. An unaccountable choice. What will they say if it is proved to be true that fully one-half of all the disease we suffer from is occasioned by people sleeping with their windows shut? An open window most nights in the year can never hurt any one. This is not to say that light is not necessary for recovery. In great cities, night air is often the best and purest air to be had in the twenty-four hours. I could better understand shutting the windows in towns during the day than during the night, for the sake of the sick.

The absence of smoke, the quiet, all tend to make night the best time for airing the patient. One of our highest medical authorities on consumption and climate, has told me that the air in London is never so good as after ten o'clock at night. Always air your room, then, from the outside air, if possible. Windows are made to open, doors are made to shut—a truth which seems extremely difficult of apprehension. Every room must be aired from without—every passage from within. But the fewer passages there are in a hospital the better."

Singular Physiological Facts.—It is impossible to think large, manly, beautiful, virtuous thoughts, while respiring in an atmosphere of stagnation and consequent corruption. People who sleep in close, ill-ventilated rooms, are forever dreaming a set of monotonous dreams, loaded with vicious pictures, and animated by strangers or demons made from the confined air. Idiots breathe superficially. They seldom respire like an intelligent mind. Timid persons inhale small quantities of air. *The coward has a narrow chest*, and he only uses the upper portions of his lungs. If a blacksmith is about to lift a heavy hammer, or strike a hard blow, he will (unthinkingly) expand and fill his lungs to their utmost capacity. Why does the strongest horse always have the broadest and deepest chest? The mind cannot expand and improve, morally or intellectually, unless the lungs be large, full, and constantly and plentifully supplied with air, fresh from the vintage of immensity. No high and magnificent conceptions can be obtained in a confined atmosphere. Mountainous air is essential to mountainous thoughts. The atmosphere of infinity is indispensable to spiritual expansion.

Do you question these statements? Are they extravagant? We challenge you to a successful contradiction. The lungs and the brain will correspond throughout in their capacities and operations. While writing these sentences, we detect a "breathless" silence ever and anon. And why? Simply because, between the appreciation of a thought and the formation of the same into suitable language, the activity of the mind is suspended by a sort of hesitation, and the respiration, as a consequence, is correspondingly suspended. This same ‑ result is brought on by an intensified effort to catch and appreciate a fine passage of poetry, or some thrilling and sublime thought evolved by an eloquent speaker. The audience is spell-bound, held "in breathless silence" literally, until the impression is perfectly received. Then respiration is deep and hearty, and, with the augmentation of muscular strength as a result of breathing a large quantity of air, there rolls out the hand-and-foot applause, so common to public assemblies. If an audience be deprived of pure air, the best speaker cannot awaken expressions of enthusiasm.

COMMON LAMENTATIONS.—What is the cry of our fast-going people? "My food does not perfectly digest!" This is the saying all over America. "My poor head aches half of the time!" So exclaim our young ladies. "My lungs are the best part of me!" Which is, unfortunately, a somewhat common affirmation. "But my liver is diseased and torpid." This is the popular complaint. "And my bowels are slow and sluggish."

Such miserable lamentations ascend from all the most fertile portions of this glorious continent. Hundreds of physicians attribute the prevalence of hepatic disorders to the conditions of the soil, water, and air; others contend that the chief cause is lurking in the constitution of man, and hold that disease is an inevitable part of this existence; but there is, fortunately, a very general interest awakened in the direction of physiological knowledge and universal improvement, and the final fruit will surely be: *the triumphant conquest of individual man over all enemies to his "bodily ease and mental tranquillity."*

Is it not worthy of particular notice, good reader, that the majority of individuals who, as invalids, incessantly complain in the department of digestion, are the most constant violators of physiological laws? They assail their stomachs, day after day, with indecent quantities of edible substances. They consign to their inward system quarts of unsubstantial stimulants, and pounds of over-nutritious foods. And yet, notwithstanding these incessantly accumulating embarrassments to a correct digestion—which necessitate an extra quantity of exercise and a greater supply of fresh air—such persons are invariably the most unphilosophical consumers of the abounding atmosphere. They breathe the most confined air at night in their unventilated bed-rooms, and during the day many of them are *too weak* (indolent?) *to take a long breath of life.* Full breathing is accomplished solely by the WILL—unless the person is about to lift a heavy weight, strike a powerful blow, or jump a chasm. Then a deep inspiration is an involuntary act, dictated and performed by *the intuition of the sympathetic ganglia.*

STRUCTURE OF THE STOMACH.—The reader is aware that the gastric secretion is designed to accomplish a separation of the solid from the fluid portions of food. The substantial particles —*i. e.*, the glutinous, fibrinous, and albuminous portions of food —are carefully selected and separated from all the liquid contents of your stomach. But do you suppose that, separate from the nerve-energy of brain-life, the stomach has any inherent power to carry forward the chemistry of digestion? Do you not know that the lining membrane of the stomach is thickly corrugated with follicles, or pits, or minute waves of the *one* substance—involutions of the mucous membrane—by which the digestive batteries of the stomach are many times multiplied in power? And do you not also know that, in addition to the muscular and semi-mechanical activities of the whole organ, the fire and life of the lungs and brain are, or should be, incessantly and fully communicated through the pneumogastric nerve to every particle of fluid and solid consigned to the interior?

RELATION OF LUNGS TO BRAIN.—To our perceptions it is too clear to require further illustration, that the vivifying fire, the soul-energy of the body, the brain-and-nerve Principle of life, is absorbed by the lungs from the boundless ocean of imponderable elements. Oxygen, so universal, is but the vehicle of heaven's divine breath. The brain is master of digestion; so also of the just distribution of strength. The stomach depends upon the brain for a supply of all forces necessary to accomplish digestion; but the brain, in its turn, is equally dependent upon the lungs for a requisite store of electric riches and vital power. The celestial elements of infinity ride straight through the lungs into the blood; thence to the great battery of all energy and digestion, the brain; which immediately distributes to each part of the body the principles of sensation, and life, and motion.

Deprive the lungs of heaven's invisible air—shut off the supply of the vivifying principles of the divine infinitude—and the whole beautiful machinery will immediately stop! The best food in the universe could give no strength, unless first baptized in *the spirit* of the atmosphere. Air is the universal blessing! It cannot be fenced in by legislative enactments; but it can be *kept out* by the ignorance or inattention of invalids. Some persons seem afraid to *expand their lungs to their utmost capacity*, lest something will break and let out the stream of life! Of course, good friend, you know that any *sudden and violent* conduct is attended with a greater percentage of risk. Begin deliberately and practice daily, therefore, and you will find that the *common air* is impregnated with an electric energy, which pervades, refreshes, quickens, and energizes every part of your physical temple.

Vital Electricity.—Your blood cannot circulate without the electric fire of the air; neither can a particle of food *strengthen* you without it. Without the living *energy* of air, which is obtained only by way of the lungs, no diet could be made universally nutritious. Salivary juice, as it pours from the little springs on either side of the cheeks and mouth, could do nothing without the air's vivifying electricity. The gastric fluid—although loaded with its inherent *pepsin*, and the acids, *lactic, hydrochloric,* &c., &c.—could accomplish nothing without a constant supply of nerve-energy. The lungs must absorb the electricity of the measureless immensity; otherwise nothing strong can occur, but death and transformation will hasten into the temple.

Suppose chemistry does establish the fact, that all food contains oxygen, hydrogen, carbon, etc.; yet it remains a problem exactly how the *heat-making* and *strength-generating* processes are maintained. If these important processes could be originated by the electro-magnetic battery, and if it were found that a *breathing* soul was not necessary to the correct performance of the digestive functions, then it would be perfectly safe for mankind to violate the Laws of Health both day and night. But the fact is that nothing vitally chemical can occur in man's body without the superintendence of the chemical principle of life in the spirit. Nitrogenous and non-nitrogenous substances, suitable for man's diet, are nothing in themselves; they must be associated with, and indorsed by, the electric fire of the brain which is supplied by means of the lungs. All the non-nitrogenized foods and medicines—starch, gum, oils, &c.—are nothing, unless empowered by the *vivifying principle* of the surrounding immensity.

Breathe, then, good reader! Take in large quantities of the divine, immaculate fire! Let every woman (and every lady, sans Fashion,) give free play to her lungs; and let every man also open his mouth to the blessings of air, ("No smoking allowed,") so that all food swallowed shall be transformed into the foundation of rosy Health, and every soul be a living fountain of gratitude and gladness. Bathe your body with pure and beautiful water; then rub your entire skin briskly with a soft flannel cloth; lastly, make passes with your hands—so that an electrical condition will exist externally, and, as a consequence, the internal surfaces be supplied with healthy *magnetism* The opposite of this condition is disease. When the surface is hot or feverish, the vital parts are negative and electrical; the exact reverse of the state of Health—when the surface is pleasantly cool with electricity, while the interior is magnetically warm with vital energy.

THE GASTRIC METHODS.—The reasons in favor of full and intelligent respiration are numerous and easily understood. Chyle is the last result of fundamental digestion. But, in itself, chyle has no power to promote growth, give strength, or repair the waste of the body. It is the successor of chyme. Chyme is manufactured from the food in the first departments of digestion. It is a pulpy mass, impregnated or charged with electricity of the vitallic kind. When it passes downwards into the second stomach, or *duodenum*, the pancreatic fluid and the bile at once combine with it, adding a positive element, by which chyme is transformed into a *milk-white* liquid (the chyle) which, with the residuum, flows steadily into and through all the small intestines.

What next? The numerous mesenteric glands, with the lacteal vessels, commence their work of forming incipient *eggs* from the chylic fluid. The unchylified portion (the residuum) meantime passes onward into the larger and lower bowels, and is thence rejected with the broken-down blood globules in the shape of bile and relative excretions. This material is wholly excrementitious. Now the thoracic "duct," so called, attracts the chyle from the lacteal passages and mesenteric glands, and pours it into a vein, which, from behind the collar-bone, discharges its contents into the positive side of the heart. Here the chyle is mixed with the negative portion of the blood. This venous blood is no more nutrient than the chyle; neither can give strength and repair waste, unless cleansed and electrified.

The Purifying Ordeal.—How is this accomplished? By means of the pure air of space! Yes, when heaven's divine breath enters the air-chambers, the chyle is converted at once into *nutritious* blood adapted to the multifarious necessities of the arterial system, and the cold venous blood is at the same moment unloaded of its death-burdens, in the form of carbonic gas and useless water. Carbon is the principal element of decay and death; yet it is essential to life, and a good conductor of electricity. This carbon is seen in the dark color of the blood. It must be disengaged and repelled from the body, or disease will ensue. The vegetable world wants the carbonic element. Death and life in the same organism!

So, therefore, the heart wisely and energetically throws both the chyle and the venous blood upon the entire responsibility of the lungs. When the invisible air is instinctively drawn into the pulmonary structures, the eternal life of the divine and infinite enters also, whereby the chyle is changed as by magic into a *constructive principle for the soul's good*, while the newly purified blood, re-baptized and confirmed in the ways of righteousness, hastens upon its mission of benevolence to all parts of the physical temple.

It is generally known that, although the element *nitrogen* remains nearly the same as to quantity, whether inspired or expired, yet the quantity of oxygen is lessened by every inhalation of air, and the quantity of carbonic acid is increased with every exhalation, all which, without argument, goes to establish the fact that human beings cannot with impunity breathe over and over the confined air of improperly ventilated apartments—that small quantities of air will not suffice to keep up the dynamic processes of beautiful health. One hundred and forty-six Englishmen were imprisoned in a room about eighteen feet square. The ventilation was insufficient, there being but two small windows, in one side, to admit the atmosphere, and the effect was very soon fully manifested. Only twenty-three of the one hundred and forty-six strong men were alive ten hours after their imprisonment in the dungeon! From this terrible circumstance the place received an appropriate epithet: "The Black Hole of Calcutta."

MORALITY OF PURE AIR.—How many superficial breathers are there, whose lungs never receive the full ventilation required? Many a human system, we think, being filled with broken-down blood globules, and other deadly impurities, may, with propriety, be styled, "The Black Hole of Calcutta!" School-houses, churches, bed-chambers, legislative halls, and every habitation, in short, occupied by organizations with lungs, should be constantly supplied with plentiful quantities of air, composed of twenty-one per cent. of oxygen to seventy-nine per cent. of nitrogen—otherwise it will be impossible for the best Doctors of Divinity to keep their congregation out of Perdition, and equally impossible for Doctors of Medicine to rescue their families and patients from the trials of private Purgatory. No true breathing *for remedial purposes* can occur unless accomplished by the WILL. It is strictly a *Pneumogastric* exercise, regulated by design. Any one acquainted with the physiology of respiration knows, that with every expansion and contraction of the lungs—or whenever the air enters and departs from the chest—many motions and changes take place in the abdominal cavity, alimentary canals, stomach, liver, diaphragm, intercostal muscles, &c., &c.

There is a deep and beautiful philosophy behind all this, which our weak and feeble Brothers and Sisters would do well to study and heed. *Food cannot impart a particle of strength independent of the lungs.* Do you believe this assertion? Do you believe that no amount of finely-prepared and costly nutrition can be nutritious, until the lungs perform their appropriate offices in the premises? It is even so, dear doubtful reader. Open the clear eyes of your Reason and see for yourself. Look straight into the breathing department, and judge whether these things be so or otherwise.

Gross matter does not, cannot, strengthen the living, vital, nervauric, immortal Principle. Your weakness is not structural. The bones are not suffering, but the *life* of them is yearning for an increase of energy; so of your internal organs—the tissues, the membranes, filaments, fibers, nerves, and muscles. These fine ponderables are destitute of the imponderable principles. You fancy that *matter* in large quantities will strengthen you. Hence you breathe little and eat much. If you should *exercise* you would of necessity *breathe more air;* then, indeed, it would seem that the food does strengthen your body; but, believe us, the facts are that the imperishable elements of strength are drawn more from the air than from the materials consigned to the stomach.

Let us look into this for a moment. It is undeniably true that the food we eat seems to undergo chemical decomposition independent of the pulmonary functions, but there is no mistake more fatal to a correct comprehension of the life-giving processes. The story is a short one. •Food is of no consequence as a strength-generating substance, until, in the form of chyle, it visits the pulmonary department and receives copulation and prolification from the electro-magnetic principles of the air. Oxygen is the royal conveyance, by which the deeper vitalizing principles drive into the constituents of chyle. As soon as a fructifying and impregnative conjunction is formed between the chyle and the air, then, and not a moment before, the food is prepared to build up and re-make the ponderable organism. If the air is impure in quality, or limited in quantity, the effect is instantly impressed upon the fluid material. That our strength is not dependent upon the amount of nutritious food we eat, is established, beyond the possibility of mistake, by the fact that persons with lung diseases, consumption, &c., usually eat far greater quantities of food than perfectly healthy individuals, who yet have forty times the volume of strength.

CONCLUSION.—We need not further amplify. The facts must be self-evident. Strength is born of the imponderable elements of immensity. The great receptive mechanism—made up of cells, blood-vessels, pneumogastric and sympathetic centers, vegetative ganglia, and bronchial tubes, ramifying in every direction—is situated in the chest. The right side is more largely supplied than the left, in order to give adequate space and action to contiguous parts and organs. The atmosphere of space, on entering this beautiful mechanism, empowers the food to supply waste and to gratify the bodily needs. Strength is the natural issue of such supply and of such gratification. Digestion is never perfect unless the respiration is full, and performed in the baptismal font of pure air, which is a vast ocean of life and energy at least fifty miles deep, and equal on all sides of the revolving globe. You will now, far more than before, understand the importance of *breathing*, (as directed,) when using the pneumogastrical cure for pulmonary and abdominal diseases. If you wish to acquire absolute strength of body, if you desire a clear and well-balanced brain, if you want a large mind and a more noble character—then, Breathe, Breathe, Breathe "the breath of life, and become a living soul."

CHAPTER X.
BLOOD, BILE, AND BOWELS.

It seems to us that the pleasures of health are beyond description. To substantiate this conviction, we refer to the stacks of medical works, to the entire catalogue of poetical eulogy, and lastly to the eloquent reflections of every intelligent invalid since the world began. The care-worn and diseased physician remembers the time "when all life's sunny hours were freshened by the breath of health." So, too, the poet, "with aspect wan and sunken eye," dreams of happy sunshine days, when the music of birds, the ringing laugh of merry children, and the romantic scenery of youthful years, kept tune to the heart-beatings of physical harmony. And thus, in short, it is with every other mortal, who, being crippled and incapacitated by disease, reflects back through the golden hours, when life's bright currents ran merrily through the Heart.

We this time use the word "Heart" in no spiritual sense. The physical organ is the everywhere-acknowledged regulator of life's magic stream. It dilates and contracts, when healthy, with equal joy and pleasure. Like a jewel hidden in the "bosom of the deep," like a bark on the trackless way of many waters, so is the visible organ "heart" in its relations to the crimson stream of life. It reflects the pleasures or the tempests of the more inward soul. The wondrous dynamics of pulsation lie deep beneath the physical structures. The principles of motion and life co-exist and work like brothers in that gentle current, the noiseless "blood."

Of the *blood* and the *heart* we have very much to write. One thousand times, no doubt, our spiritual eyes have peered into the secrets of the life-fluid. Its constitution, its mission, its beautiful operations throughout the whole physical mechanism, and lastly, its *diseases*, have painted, with unrivaled pencil, many most important truths upon our understanding. A few of these we present, with the hope that some reader may receive the truth and be thereby directed into ways of gladsome health.

First: The blood is manufactured out of materials consigned to the stomach. The physiology of this process is exceedingly beautiful in health, but we will not dwell upon it.

Second: Digestion is a marvel in the chemical laboratory of life. In health the mind is *unconscious* of this many-sided process. The mucous membranes co-operate with the muscular tissues; fluids and ethers, time and temperature, acids and alkalies, reciprocate each with the other throughout; so that, in health, the most sensitive mind can realize nothing but pleasure and the accumulation of abundant power to execute the duties of life. The magnetic fluid, termed "gastric juice," receiving its subtile energies from the brain, through the great sympathetic nerve, can convert any soluble substance into a limpid nutriment. This is the *chyme*, which, settling into the duodenum, soon mingles with a discriminating fluid, termed *pancreatic juice;* and the bile, with its negative qualities coming in to aid the processes of separation, soon ultimates the food into a fine fluid (chyle) which is the material for the immediate production of blood.

Third: Let no one suppose that the blood is red or blue in the beginning. It is clear and odorless as pure milk, with but little coloring properties, when absorbed by the hair-vessels that line the small intestines. At first the blood is composed of innumerable *eggs*, which are originated in the *lacteal* membranes. These vessels and minute membranes constitute a perfect *ovarium*, wherein the globules of the blood are primarily formed, and from whence they are subsequently detached ; when they drop into the flowing currents and thence float off into the general circulation. We do not give details, because they are deemed unimportant for the purposes of this chapter, which is to indicate a few facts in the cause and cure of disease.

FOURTH : The unnumbered spherical bodies or globules are each a center of *life* to the individual. His blood is a moving miniature sea of oval forms, of countless nuclei, of points and pivots, upon which all the life-wheels turn and spin the web of spirit. Each sanguinous *egg* is also a center of vitality for the perpetuation of the race. Let the physical eye inspect this ovarium, and let the chemist break its eggs, and classify their contents, and he will speak (1) of *red* globules, (2) of *Lymph* globules, (3) of *Chylic* globules, and say that the composition of *healthy* blood consists of so much *serum*, so much *fibrine*, and so much *albumen:* all which, by further analyzation, yield many mysterious properties—sulphates, phosphates, carbonates, chlorides, peroxides, &c., &c.—but the great internal facts and laws, which are fundamental to the existence and healthy performance of blood, remain wrapped in folds upon folds of materialism. In proof of this we refer to the custom, not yet extinct among best educated physicians, of *blood-letting*. What can more clearly establish their utter ignorance in respect to the blood's internal nature and mission in the economy ?

FIFTH : Not attempting a line of detail concerning the *modus operandi* of the circulation, showing how respiration gives color and vitality to the heart's fluid, we proceed at once to inquire,

What is Bile? What is its office in the digestive mechanism, and how does it originate so many diseases? "Bile" is a bitter liquid, of a brownish green, very negative, and pervaded with an acid mucous. What labor does it perform? According to our examinations, we affirm that the hepatic bile, which flows from the liver into the duodenum, performs the office of separating the mucous and innutritious particles from those which are suitable for becoming chyle and globules. Many troublesome disorders originate just at this point, such as *sick head, gastritis, indigestion, dyspepsia*, besides a multitude of symptoms which indicate discord of more or less extent and severity. That form of bile which belongs to the oblong receptacle, termed the gall-bladder, performs the office of still more liquifying the contents of the stomach and duodenum. It is composed of the excrementitious portion of the hepatic secretions, which the receptacle discharges through the "cystic duct" into the bowels. Hepatic bile is highly carbonaceous until it impregnates the fluids of the middle stomach, when it becomes cold, indigestible, and wholly excrementitious.

SIXTH: But there is one thing very remarkable; this excretory and innutritious portion of the biliary fluid never passes off the bowels unless every part of the intestinal machinery is in perfect and prompt working condition. The cause of this fact is not explained by physicians. The usual term for the effect that follows is, "biliousness." The victim is sleepy, headachy, stretchy, chilly, yawny, and "don't feel very well." It is known that bile supplies carbon to all the matter destined for blood globules. But the pancreatic fluid is a powerful ally in the work of separating dense from rare properties, and in preparing every suitable particle for chylification, into which the great sympathetic nerve is perpetually discharging streams of magnetic energy. The pneumogastric nerve is most affected when the mind is fixed upon any subject too soon after eating; for through it an electrical influence is steadily imparted to the mucous membrane of the entire digestive system, a process which too severe moral disturbance or study very soon impairs and arrests. But bile is something more than all we have described; *it is broken-down blood globules;* it is the refuse material of the entire *ovarium;* it is the *mud* of the waters of life, the *husks* of the corn, the *shells* of the blood eggs; and this is the reason why the whole body "travails and groans in pain" whenever such *debris*, by etherization or absorption, are taken up into the circulation instead of leaving the temple, at the appropriate time, by the natural avenues.

The remedy adapted to one person may be non-effective in the very next patient. A few simple rules are invariably efficacious, both as preventives and curatives; such as regular meals, proper mastication, not much fat or gravy, no heating stimulants, no cakes or pastries, and punctual attention to *every* natural function. But when the debridation of the *old blood* regurgitates into the circulation, there is, then, no such thing as wholly removing it by *dieting, bathing, brown-breading*, nor by any other gentle method popular with the "no-medicine fraternity." Certain temperaments may, it is true, succeed by a persistent course of *diets* and *bathings;* but the great mass of mankind would, by such means, fail both in strength and in the object of their exertions.

For this reason we shall prescribe preparations, very simple in themselves, which cannot fail in aiding the process of *chylification;* and thus, consequently, facilitate the escape of the *broken-down blood globules* (or excrementitious bile matter) from the circulatory and digestive systems. Let every so-called *bilious* or *jaundiced* person, inclined to symptoms already mentioned, who is laboring with other sensations

8

characteristic of over much biliary fluid, mix one tea-spoonful of powdered willow charcoal in a wine glass of pure Holland gin, or water, and swallow the whole just before every third dinner, or twice a week.

In severe cases, where the system is subject to great depression for days together, with weakness and loss of appetite, &c., take the same preparation, with same quantity of rhubarb, just previous to every dinner, and continue the potions for eight or ten days.

Now abstain from all medicines of every kind for a number of days. If, then, the bile and bowels continue to indicate derangement, repeat the course as before; being meantime extremely temperate in regard to rich food, and avoiding exposure to chilly atmospheres. There are more searching and more simple remedies, but we will begin with charcoal and rhubarb. The time, however, will surely come when men shall scorn all medicine, all nauseous compounds of both doctor and priest, and, breathing the sweet air of heaven, will sing, "The ways of wisdom are ways of pleasantness, and all her paths are paths of peace."

CHAPTER XI.

THE PHILOSOPHY OF HUMAN MAGNETISM.

THERE is a very common superstition among popular medical men of the antediluvian school, that the intellectual phenomena of magnetism (or "mesmerism") are the concomitants of hysterical states of the nervous system. Old line doctors attempt to transcend the otherwise insurmountable difficulties of somnambulism and clairvoyance, by the assumption of imposture, or else by charging the mental manifestations to nervous or cataleptic conditions of body and brain. It is, however, very generally believed that the majority of diplomatized physicians are well supplied with *ignorance* concerning many of the most vital processes of the physical organization. Chemistry has recently enriched the physician's understanding of physiological phenomena. But chemistry does not unravel to his mind the wondrous dynamics of the feeling and thinking principles, which animate and govern the perfect and beautiful organisms of men and women. The mental and spiritual phenomena of magnetism are yet new to most physicians, and we do not, therefore, expect anything else from them than expressions of professional prejudices, emphasized by strong marks of dogmatic denunciation. But there is, here and there, a broad-hearted and knowledge-loving physician, who is capable of putting a rational question, with an honest incredulity; and who, consequently, is ever ready to exchange his learned errors for new truths—is willing to make progress in scientific facts, and to unfurl the "Union" banner of free thought and unlimited investigation.

But it is not the design of this chapter to construct an argument for the genuineness of magneto-mental phenomena. We can scarcely believe that such an argument is demanded by the so-called scientific of the age, and yet we know that no class is more in the *rear* of advanced discoveries than the graduates of our institutions of learning. Many of our best students in Medicine are unable to solve the first group of magnetic phenomena. They treat the alleged facts as obviously incredible and impossible, and so permit themselves to be sufficiently *illogical* to reject the facts, and sometimes enough *uncivil* to insult the "hewers of wood and drawers of water" who have the audacity to present such phenomena for scientific examination.

The colleges and the churches are proverbially *behind* in the essentials of knowledge and civilization. The unscientific "people," the non-professional observers of Nature, and the clear-eyed, matronly nurses of the sick, are the unconscious champions of scientific progress. After these, like a loaded omnibus behind the laboring horses, come the respectable host of physicians and clergymen—riding, and enjoying themselves luxuriously, in the cushioned chairs of our Collegiate and Evangelical Institutions. Millions on millions of human beings, as well as creatures in lower grades of animation, breathed the "breath of life" all unconscious of science—unmindful of that chemical knowledge which would explain the constitution of atmosphere, and reveal the proportions of oxygen and nitrogen to the thoughtless multitude.

So in every other respect. The people intuitively illustrate the essential facts of science for centuries in advance of the accurate knowledge of the schools. In human magnetism this remark is emphatically true—" the people," with little or no education, are familiar with its essential facts, and have practiced the principles of the science long eras before the Colleges reflect the first ray of light on the subject. But when the Colleges and Academies adopt the new science, and the professors venture to instruct their classes in the fundamental principles of the phenomena, then behold the supercilious pomposity of the learned dignitaries, who unblushingly inform the children of the populace that Science has developed the new facts and principles. The truth is, " Science " is nothing more than the systematic observation and orderly arrangement of those natural *facts* and superficial *causes*, which have for hundreds of centuries been *common and familiar to the inhabitants of every country*. It is, therefore, no disadvantage to any experience or philosophy, to say that it is not yet accepted and inculcated by talented men in high places. Because, as we have shown in preceding remarks, the knowledge of Colleges, and the theology of the Churches, are reflections of the facts and discoveries of the Past. "The people," on the contrary, without education, are masters of realities and principles not yet " dreamed of " in the brains of our academical professors and evangelical teachers.

THE SOURCE OF MAGNETISM.—We employ the term " Magnetism " in its broadest sense—signifying the principle by which one object is enabled to attract, repel, and influence another. The source of this principle is SOUL. Crystals, various mineral bodies, plants, trees, fish, birds, animals, human beings—each and all are endowed with the magnetic principle, because each and all are endowed with a SOUL, which is the mystic life of boundless Nature, upwelling and ever-flowing from the inexhaustible Fountain of the Great First Cause. (Students and readers who are intellectually acquainted with the Harmonial

Philosophy, will not confound *Soul* with Spirit.) The term "Soul" is here used to signify that harmonious combination of the principles of Motion, Life, and Sensation, which *move*, and *warm*, and *perfect* the physical organization. Stones, trees, animals, men, contain these principles; the latter in a high degree of development, while in the former, the principles are comparatively dormant and unfledged. Each natural body of matter is differently capacitated; hence, also, it is differently supplied with the Soul-principles. The consequence of this difference is a *magnetic polarity* between one body and another throughout the entire domain of Nature. And the consequence of this universal *polarity* is the evolution and manifestation of all the physical motions and mental phenomena, known or unknown to science.

Facts Illustrative of Magnetic Polarity.—The common magnet, as every reader knows, is at once positive and negative. That is, the life of the metallic body makes two manifestations at the same moment. It will attract a negative substance, and repel that which is positive to it. The positive pole is charged with negative power, and the negative pole with positive power, and the manifestations of the magnetic principle correspond to these facts. The seed of a plant is negative to the magnetic heat of the sun; consequently, the properties of the seed, if sown in good ground, leap up toward the magnet, as the needle points to the pole. This explains the growth of vegetation. Thus the near relationship of magnetism and electricity is demonstrated. They mutually attract and mutually repel each other. Look at the common electro-magnetic battery. If the electric current be permitted to traverse the coil of wire, it will convert the rod of iron, placed in the center, into a powerful magnet; and this, in its turn, will set in motion a powerful current of electricity, as it were, by way of compensation.

The human body is constituted on the same system of polarity. Man is polarized from side to side, from end to end, from centers to the surfaces. His nervous system is a network of polarities. From his inmost organic centers to the glands of the brain, and from the brain-centers to the extremit͏̈ of every nerve, he is a perfect battery of magnetic and electrical potencies. The entire left side, from brain to toes, is negative. The left-side emanations are therefore tranquil and attractive; while from the right side, which is positive, the emanations are powerfully repellant. Hence, man repels, and works, and destroys, with his right side, right arm, right hand, right leg, right foot, and brain; while, with the corresponding parts and members of the *left* side and brain, he attracts, and subdues, and magnetizes whatever he is adapted to affect. The right side of the brain is frequently unimpressible, while the left side may be easily overcome and paralyzed by the magnetic principle. The right eye, in healthy persons, is the keenest and best; while the left eye is capable of more pleasurable vision. The left eye of a susceptible person will, for this reason, more readily discern the *colors* of a substance. The location, the size, the weight, and the distance of a body are quickest determined by the right eye. If the reader doubts these statements, let him experiment with his eyes and senses. Close your left eye and look at the leaf of a plant; then reverse the method, and your left eye will soon begin to see rays of light, which your right eye cannot discover. In like manner, if you have much susceptibility, your left hand will detect *heat* in substances which are *cool* to your right hand, and the reverse is equally true, only frequently practice, with care and discrimination. For these reasons the right hands of man and woman are attractive to each other, while, many times, the hands of the same sex are mutually repellant and unwholesome.

Clairvoyants can detect the emanations of the different centers by the *color*, which is natural to polarized principles.

THE PRACTICAL WORKINGS OF THESE FORCES.—The source of the magnetic force is the SOUL, and the effect of the power corresponds to its source—that is, the power is lodged in the SOUL of the subject, and the manifestations are, therefore, more psychological than physical. We will suppose, for illustration, that two healthy persons seat themselves, (as in figure No. 1,) to try the magnetic experiment. They naturally face each other, which is in philosophical harmony with the polarities of the magnetic principles; that is, the *right* side of the operator is presented to the *left* side of the subject. Previous to the experiment, we will suppose each person to be in separate and distinct states, wholly independent of each other, with respect to sympathies and antipathies; which important fact the artist has attempted to illustrate, by the separate oval dotted lines surrounding each individual.

Fig. 1.

The experiment is now to commence. Could your mental eyes be suddenly opened, as is the case with clairvoyants, you would behold a wondrous exemplification of a great general law of Nature. The right side of the two persons would glow with flame-like emanations. At first a gray colored light would stream faintly from the right side of the brain, and thence downwise to the ends of the right hand and foot. The natural forces of brain, and lungs, and heart, and stomach, would present a fiery appearance, but variegated with many colors, like those of the rainbow, or like the electrical emanations of millions of differently constituted plants and flowers. The fingers would seem to glow like tapers in a dark night. In short, the form of each person would seem to step out of darkness, and to be filled with effulgence the most beautiful and attractive.

We are supposing, remember, that the operator and subject in our experiment are magnetically related to each other, so that there can be no failure in the progressive application of the principles under consideration. The wonderful and complex nervous system of man is a complete helix, a coil of wire, which communicates electricity to the brain, which is *the magnet*, or central power, of the organization; and the compensating process, as with the electro-battery, goes on in the shape of centrifugal currents of nerve-life, (a finer electricity,) which the brain discharges through the pneumogastric and sympathetic nerves to all parts of the temple. We cannot now stop to detail the beautiful facts of this process, but may on some future occasion.

In accord with the magnetic law, we next observe that the brain and body of the operator become one o'ermastering positive power, to which, without resistance, the subject surrenders himself, both physically and mentally, and the resulting manifestations are what is usually denominated "psychological." The partial blending of the magnetic spheres of the twain, is illustrated by the interlocking of the dotted lines, (see figure No. 2,) showing that subject and operator are magnetically more closely related as members of one body. In this condi-

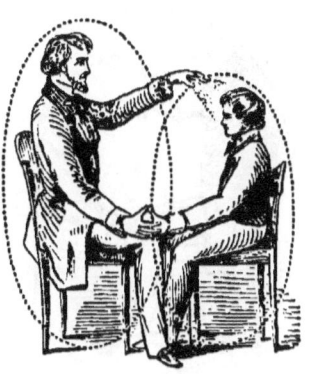

Fig. 2.

tion, the operator's SOUL is the center of attraction. The subject's attention is identical with the operator's. By the mere exercise of fancy, without the least mandate of will, the operator may image his thoughts upon the subject's brain. He may cause him to drink *wine* from a glass of pure water; to hear the roaring of cannon and clashing of weapons on the battle-field; to feel the strength of a giant; to catch fish in an imaginary stream on the carpet at his feet; to weep the tears of sorrow at sufferings purely fictitious; to pray for forgiveness at the throne of an implacable potentate; and lastly, to forget his own individuality, and take on the feelings and exhibit the striking characteristics of the operator, or of any one whom the operator has the intellectual power clearly to shadow forth in the positive odyllic light of his own mind. This psychological law lies at the bottom of all that class of so-called " spiritual phenomena," wherein, to the observer, it seems that the spirit or mind of the medium has vacated its temple in order to give a foreign intelligence an opportunity of manifestation.

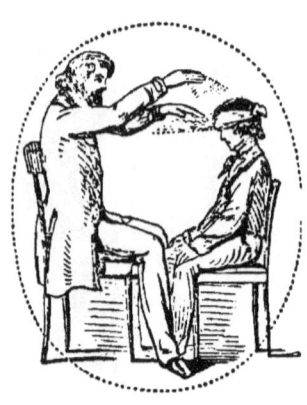

Fig. 3.

One step further on in this magnetic career will be followed by the *complete blending* of the vital and mental spheres, (as illustrated by figure No. 3,) in which case are exhibited all those mysterious and glorious phenomena termed " Somnambulism," " clairvoyance," " spirit seeing," &c. The extent of man's capacity in this peculiar state is not easily measured. The subject is no longer psychological or sympathetic. The condition is most favorable to very high perceptions of natural truths. The clairvoyant is capable of medical examinations; also, as a "sensitive," of testing the positive and negative qualities and polarities of crystals, metals, medicines, waters, bodies, &c. Some persons there are who seem to be born with the last-named gift, and yet without the first symptoms of natural clairvoyance. Reichenbach terms such persons "sensitives," because they are *clear-feelers* rather than *clear-seers*, or clairvoyants. The German philosopher says: "Suppose, now, that there were a vein of lead, copper ore, or red silver ore, not far below the surface, as they are often found; if a high sensitive were to walk over them, with attention, he would feel them and be able to tell their position. Stone-coal exercises an odic influence, different from that of sandstone and slate, in which it is found. If the sensitive has paid attention, beforehand, to the sensations which coal causes, he will readily recognize them when he approaches a vein of coal. Non-sensitive men will not be able to feel anything, but the high sensitive will be able to say with certainty, Here or there, this or that mineral may be found in the earth; and, by digging, proof will be found of the correctness of the assertion, which appears so much the more wonderful from the fact that the treasure finder can give no satisfactory explanation of the manner in which he made his discoveries. The marvel is now exposed: it is a purely physical effect of the odic force on the human nerves; it works like a dark sense, of which we can give no explanation; and a multitude of instinctive actions among brutes will find their explanations in the same way. And now you have the whole secret of the divining-rod; not of the rod in its literal sense, and of its rising, falling, and turning; these were only the hocus-pocus for the inquisitive crowd, who would not be satisfied until they could see something.

"You perceive from this how great the practical importance of sensitiveness, and what a career it is destined to have. These sensitives and somnambulists will soon be sought and counted as the benefactors of their neighborhoods and countries. To mining, this discovery promises an extraordinary development, and this not only by the discovery of new beds of ore, but also for the running of their shafts underground, when the stratum eludes the miner." We must here express our conviction, that the pursuit of subterranean knowledge will not promote the development or happiness of the "sensitive," or clairvoyant, who so employs his spiritual power.

Fig. 4.

In order to exhibit the full course of the magnetic experiment, we introduce the ultimate state, called the *Superior Condition.* The dotted oval lines, which illustrate the magnetic rapport of the operator and subject, show (see figure No. 4) that the twain are related only through the vital powers and processes. The *brain* is now completely emancipated from the pre-existing magnetic thraldom, and consequently the mind of the clairvoyant is *independent* of all surrounding circumstances. Once for all, let us remark, that the magnetic process will not guarantee to *every* person these succeeding phenomena, any more than going through college will insure to every scholar the development of a Shakspeare, a Bacon, or a Plato. Favorable proclivities and organic qualifications precede the production of the mental phenomena. Neither will it be possible for the magnetic *sleep* to succeed the passes in every case; all these effects follow in a train of favorable causes and predispositions, or they do not at all appear. And yet, in justice to the endowments of our common humanity, it is but simple truth to say that there exist in every person, of every nation, the germs and faculties of all the grandest powers ever exhibited by any human mind. Their development and fruition are certain in the march of Time through the ages.

MAGNETISM AS A MEDICINE.—Having briefly sketched the action and mental effects of the magnetic principles, it is now expedient to conclude our remarks in behalf of the sick and suffering. The human body, in its normal and healthy condition, is endowed with every requisite power. But by ignorant and negligent treatment, the natural vital forces lose their just equilibrium, and the effects and consequences are soon visible in material prostrations, in severe pains, or in silent and insensible decomposition. What physicians term "nervous influence" is really nothing but the magnetic and electric life of the interior SOUL. Animals, including men, have these magnetic endowments; and the principles of vital action, in both the human and animal kingdoms, are exactly and universally identical. A loss of vital action is nothing but a loss of *balance* between inherent forces, which are positive and negative, or magnetic and electrical. And yet we do not hold that the currents generated by the metallic or mineral battery can ever be made to act as a substitute, because the principles of SOUL-*life* are as much more fine than atmospheric electricity, as the latter is finer and more delicate than the gross and turbulent water of our lakes.

Therefore we recommend the judicious use of human magnetism in nearly all cases of disease—especially the use of your own magnetic energy on different parts of your own body! Your left side can treat your right side; your right side can magnetize your left side; your vital centers can give the surfaces a thorough magnetic sweating; your hands will do the bidding of your brain; and your brain will act obediently to the commands of a well-ordered judgment. "Ah!" you despairingly exclaim, "I've tried the experiment, and cannot succeed." We reply: "You do not succeed for the same reason that a boy cannot swim, or skate, or accomplish anything correctly, until the art thereof is fully and systematically acquired." We prescribe different remedies merely as *palliatives* and *aids* to your final redemption from disease, and from the fear of death, but *the radical remedy is still within your own individual organization.*

The therapeutic influence of magnetism may be exerted in various ways, differing in every case with the temperament and the nature of the disease, and for this reason we do not attempt, in this chapter, to specify methods. It should, however, be borne in mind, that to practice magnetism successfully, (as the distinguished M. De Puysegur said,) "You must have an active WILL *to do good, a firm faith in your power, and an active confidence in employing it.*" Magnetism is a useful, a spiritualizing, and a sublime agent of energy and health. It is the all-pervading *sympathy* which connects us with the absolute condition and sufferings of our fellow men. Owing to the delicacy and sublime uses of the magnetic power, it is susceptible of remarkable *mis*-applications, much to the annoyance, perhaps injury, of both the operator and subject. Prof. William Gregory, late of the University at Edinburg, said: "I have been informed, on perfectly good authority, of the case of a lady, highly susceptible to the magnetic influence, who could never be magnetized if a certain person were present; and I know another lady, who is easily and pleasantly magnetized by one person, while the magnetic influence of a third individual is to her insupportable."

The same excellent authority says: "Another class of failures depends on a different cause; I mean, the prevailing fallacy, that all cases of animal magnetism, in their different stages, exhibit precisely the same phenomena; that is, that if we have seen, or read of, a case, in which the various stages of the state of somnambulism have each exhibited the principal phenomena peculiar to such stage, the next case or cases must, of necessity, present the same facts, and in the same order. This fallacy is nearly universal, and the consequence is, that many persons, who have seen, or heard of (for example) thought-reading, or clairvoyance in any other form, in one case, cannot imagine that these phenomena may be absent in another. They clamor for what they have seen before; the exhibitor rashly tries to produce it; but the subject is an inferior one, or in a different stage, and entirely fails to realize the expectations so ignorantly formed. This, however, would be nothing, were it not that the failure is seized on by many as a proof of imposture. It proves, however, only this: that the spectators were mistaken in expecting the same results in every case, and the exhibitor entirely wrong in attempting to gratify them. Every case must be studied for itself, and, although certain general laws apply to all cases, yet the variety in the details, both as to their nature and degree, is infinite.

"Not only do different subjects differ in the nature of the phenomena they exhibit, as, for example, when they can only be got into different stages of the somnambulistic state, each persisting in his own stage, but, even in the phenomena of one stage alone, the same variety is observed. Thus, in the lucid, or clairvoyant stage or state, some are utterly insensible to all sounds save the voice of their magnetizer; others hear every sound, often with increased acuteness. Some will only answer the magnetizer, or those placed by him *en rapport* with them; others will answer questions put by any one. Some retain their sense of identity, others lose it. Some require contact with the person or thing to be observed, others do not. Some see their own frame, in all its minutest details, as well as the bodily state of other persons; others see nothing of all this. Some possess vision at a distance; others are devoid of it. Some can read closed letters, or letters shut up in a box, or mottoes inclosed in nuts; others fail entirely to do this, while they can, perhaps, read our inmost thoughts, a feat which, possibly, the letter-readers may not be able to accomplish."

We have now given you the general principles of the magnetic medicine treasured up in the organs and brain-centers of your own individuality. An inflammation is a *positive* condition of an organ or part; therefore, apply your positive hand and WILL to it. Why? Because two positives repel, and your hand, being a healthy positive, will scatter the inflammation, which is an *un*healthy positive, and thus establish the natural equilibrium. Your brain is loaded with blood! Not so. Your mental magnet is surcharged and overstocked with vitalic currents—which should be engaged in other parts of your economy—and thus the dependent blood is not floated off. Some doctors will bleed an apoplectic patient. This method is absurd. No man's system ever generates more blood than it needs for its own private use. But it is possible, nay, easy, for the magnetic potencies to be thrown out of *balance*, giving rise to co-ordinate symptoms of *excess* in one place, and of *deficiency* in another; the remedy in all cases being the same, viz: a restoration of the magnetic equilibrium, between foot and brain, between stomach and liver, between heart and lungs, between spleen and kidneys, and the inevitable consequence will be perfect HEALTH. And now may our Father God and Mother Nature, who are always in supernal harmony, save all the *sick*, with an everlasting salvation.

Magnetic Processes.—There are various methods practiced by different magnetizers; but (says Morley, in his pamphlet on the "Elements of Animal Magnetism:") we think the following preferable:

"If you wish to put a person into the magnetic sleep, cause him to sit as easy as possible in an easy chair, with his head reclined back, and require him to be perfectly quiet; sit down before him, place your knees beside his; then take his thumbs in such a manner that the inside of your thumbs will touch the inside of his. Concentrate your attention, and will him to sleep; after holding him thus about ten minutes, slowly raise your hands, with the palms turned outward, to his head, then, turning the palms inward, let them descend to his shoulders, and let them remain there five minutes; then let your hands descend, with the fingers pointed toward the arms, at the distance of two or three inches from them to the extremities of his fingers; let your hands then ascend, sweeping them off to the right and to the left, to their extent, palms outward, as before; raise them as high as the head; then descend, as before; thus continue from five to ten minutes, and lastly, lay the right hand upon the pit of the stomach. Remember that unless you keep your attention fixed, your will steady and unwavering, your efforts will be in vain. The operation is principally intellectual; many make no use of the manipulations, and produce all the effects by the mere energy of the will, at a distance from the patient; but still, the movements of the hands give some assistance in producing the magnetic current; the downward motions are magnetic, the upward are not. Some persons are much more susceptible to the magnetic influence than others; hence some require a longer time in being put into the magnetic sleep than others; in some cases the processes are shortened, in others they must be lengthened. There are some persons upon whom magnetism has no sensible effects. Another very successful method is, to take the patient by one hand, and place the other hand on the head, and exert the will, as in the preceding case.

"But a comparative few that are put into magnetic sleep become somnambulists. If a person in this sleep will answer the questions of his magnetizer, he is in somnambulism. To awaken the patient from magnetic sleep, make upward motions with your hands before his face, willing him to awake, and he immediately awakes.

"The magnetic sleep is highly restorative, and always should be resorted to when the complaint is general; but when there is simply a local pain or disease, there is no necessity for it. For headache, place your hand upon the part affected, and exercise a constant and benevolent desire to relieve pain; and, after holding it there a few minutes, pass it lightly over the head from right to left; if the pain is occasioned by the stomach, next place your hand on it, and proceed as with the head. If the headache is accompanied with cold feet, after holding the hand on the head for a short time, draw the hands slowly from the head downwards, along the sides, to the knees; soon the head will be relieved, and the feet become warm. If the pain has existed for years, it is chronic, and must have a prolonged treatment.

"In rheumatism, if local, place your hand where pain is felt hold it for fifteen or twenty minutes, then pass your hand lightly to the extremity of the feet, and thus continue for ten minutes; but if the limbs are generally affected, make passes at a short distance from them to their extremities, for an hour or more; if the disease is chronic, repeat the operation daily until the relief is complete; and so of every chronic disease. Says Deleuze, 'I have seen a fit of the gout, so violent that the patient could not put his feet to the earth, relieved by one sitting and cured by three, and the pains have not returned for eighteen months. I have also seen a somnambulist in fifteen days cure her magnetizer, who for a long time suffered with the gout in the knees and feet. For this purpose she merely employed passes along the legs, continuing them each day for a quarter of an hour. When the gout has mounted to the head or chest, magnetism readily brings it down to the feet, and then draws it off at the extremities.'

"We mean by pass, simply passing the hands or moving them as we have stated.

"For toothache hold the hand on the part affected for a few minutes, then pass the ends of the fingers slightly over the cheek from right to left.

"In biles, magnetize when the inflammation begins.

"For a felon, make passes along the arm as far as the extremity of the finger, and then concentrate the action, and then draw it off from the end.

"It is not pretended that magnetism cures all diseases; some are beyond its reach; but it is a valuable auxiliary of medicine, and every physician should be familiar with its principles; and a general knowledge of them would relieve many of the ills of life, and preserve multitudes from untimely graves. Says Baron Dupotet, 'The value of such a discovery as animal magnetism is to be estimated, not by the evils to which its unskillful application may give rise, but by the positive good which may be derived from it. Already we have seen that during the state of magnetic insensibility, the most painful surgical operations may be performed and the patient remain the whole time in a state of perfect unconsciousness. Is this not a boon to suffering humanity? This is not all; the most obstinate and painful chronic diseases have been relieved and perfectly cured by its application. It was the successful treatment and cure of diseases which had notoriously resisted every other remedy, which compelled the rudest and most inveterate of our antagonists to recognize the influence of magnetism; and when these facts were demonstrated beyond all reasonable controversy, it remained for them to seek in the umbrage of their imagination the solution of the mystery. In epilepsy, hysteria, neuralgia, chronic rheumatism, headache, I know of no remedy so immediate and availing. How often have I seen the victim of pain writhing in the most acute agony, sink under its influence into a state of the most placid composure! How often have I heard thanksgivings and prayers breathed in gratitude to the Creator for the relief which the afflicted have hereby experienced! At Groningen, a girl nineteen years old was suffering under hysterical spasms, which sometimes continued forty-eight hours; after being magnetized half an hour a day for three weeks, she recovered. A lady residing in London, after a violent attack of fever, under which she was suffering in December and January last, was affected by convulsions of every kind, but mostly by fainting, which often lasted two hours, and it was difficult to bring her to herself. I was present one day when the fainting was coming on, and tried to make application of magnetism; I had scarcely begun to operate when she quickly recovered from the fainting, as though she had been awakened from a dream, and from that moment she gradually recovered.' Says Dr. Elliotson, of London, one of the most eminent physicians in the British empire, 'I know of no certain cure for epilepsy but magnetism; I have cured several by it.' Says Baron Dupotet, 'In many acute diseases, medicine should be used with magnetism.'

"To cure a person of any bad habit, as intemperance, he must be put into the magnetic sleep, and then the magnetizer must WILL with energy that the least participation in intoxicat ing drinks, snuff, tobacco, opium, or whatever it may be, should cause *nausea*, and he will be forever unable to partake of the interdicted articles; unless, in another magnetic sleep, the magnetizer should remove the interdiction. It may be equally well applied to anger, revenge, and every evil passion. Hence, the philanthropist, by a practical knowledge of this agent, has his means of relieving suffering humanity increased a thousand fold; and many frightful maladies will take their flight, before its bright rays, from our globe.*

"CAUTIONS FOR OPERATORS.—Says Baron Dupotet: 'I am anxious to impress on the minds of those who may feel inclined to try the experiment, that the operation is not always unattended with danger; for I have known instances of many, who, in endeavoring to induce the magnetic phenomena, have placed themselves in a very painful position, and the person operated on in a very alarming state. Of course, animal magnetism, like every other science, has its own laws, and these should be diligently studied, before any individual attempts to practice it. M. de S. C., a retired officer, having heard a vague report of animal magnetism, attempted to make the experiment upon his own daughter, although she complained of no illness. He merely wished to ascertain whether he could make her feel the magnetic sensations. With this view, and without being aware of the extent of the mischief he was provoking, he laid his hand on the stomach of his daughter and obeyed the magnetic injunctions. After a few moments of magnetization, she experienced spasmodic attacks, and shortly was seized with violent convulsions; and her father, not knowing how to calm them, only increased their intensity, and she thus remained for a week.'

"Says M. de Puysegur: 'A young lady of Nantes, of

* A person cannot be magnetized when under the influence of any stimulating drink, food, or any excitement.

distinguished birth, when on a visit to her relative, the Marquis de B., was indulging, with the rest of the company, in passing sundry jokes upon magnetism. Her uncle, M. de B., who outstepped, by his sarcastic remarks, every one present, and was gesticulating with great freedom, began to direct his pretended influence upon his niece, when they both set about magnetizing each other as fast as they could. At first, the young lady laughed very heartily, but it was soon discovered that this laughter was anything but natural; and she was gradually losing her reason; she followed her magnetizer everywhere, and yielded to his sole influence. The spectators attempted to separate them, but this only provoked dreadful convulsions. Her magnetizer felt extraordinary sensations; the lady remained in that alarming state several days.'

"But if convulsions do occur, the magnetizer, by being calm and firm, can soon quell them, by making passes at a short distance from the patient, and directing the energy of his will to soothing or calming them. An experienced magnetizer rarely ever induces convulsions, and if he should, he can speedily remove them. In magnetizing for the relief of any local pain, there is no danger; so that any one can attempt it with impunity. In nearly all cases where there have been convulsions, they have occurred when attempts were made, through mere curiosity, to excite the magnetic phenomena. Women can magnetize equally as well as men; all nurses should be magnetizers.

"Magnetic Treatment of Insanity.—Physicians are often ignorant of its cause. There are portions of the brain that cannot bear the least pressure without derangement or fits. The organs of destructiveness and combativeness are most easily affected. When there is an equilibrium of the circulation of the blood, there will be no derangement. Lack of circulation is the first cause of insanity; then a portion of the serum does not become blood. A bruise becomes bad blood, and pleurisy is produced by thick blood.

"Different organs become deranged, as eventuality, constructiveness, secretiveness, and acquisitiveness—when the last-named organ and color are deranged, the person thus affected will steal articles of a certain color. The physician's first object should be to ascertain what organ is affected. If mirthfulness is disordered, excite veneration. Try to draw the surplus blood from the brain toward the extremities by magnetic passes.

"Insane persons ought never to be opposed. Follow them in their views, as if they were sane, by small portions, in due season. Examine the patient's hands and feet; when they are warm, and animal heat is equalized with moderate perspiration, and the system is open, the health of the person is good

"Tranquilize the patient by dieting, and not permit him to eat food that makes blood. The following articles are appropriate, namely: crackers, rice, and molasses, and avoid stimulants. Palsy is produced by a similar cause as insanity. Never let the insane know that you think them insane, as it makes them worse; and also eyeing them with suspicion does the same. The reason why their best friends prove their worst enemies is, because they eye them closely, which horrifies the insane, and increases their malady, and begets in them extreme hatred toward their friends. Insane persons should be talked with as if they were sane and rational. They ought to see frequent change of scenery, the oftener the better; and, in extreme cases, let pictures in the room be changed hourly.

"Persons that become insane by fixing their thoughts constantly on one thing, are hard to cure. Any person confined in a white glass globe would become insane in six hours. Long and intense thinking *on one subject* will render any person more or less insane.

"The doctrine of Election causes more insanity than any other one subject, because it leads its believers to doubt and melancholy, and finally to despair. Universalists are rarely insane, as they are buoyed up by hope, and are often cheerful. All insane persons are costive. Typhus fever is a species of insanity. The nerves of voluntary and involuntary motion are opposite; if the one class are unusually active, the others are proportionally inactive.

"In addition to the other remedies, give a tincture of Cayenne pepper and alcohol, and use the warm bath, with friction, by rubbing the patient with a wet woolen cloth.

"If the foregoing treatment was observed in the insane hospitals, in one week three-fourths of the patients would be cured."

CHAPTER XII.
INDIVIDUAL RESPONSIBILITY.

It is clear to a demonstration that "What we shall be doth not yet appear." By interior investigation we arrive at the ennobling conclusion that, in the future of individual progress, man's innermost is destined to become *uppermost* in the sphere of physical circumstances. We behold the consoling truth that the human spirit is constituted upon the principle of mastership and self-sovereignty. Conceiving thus, we are well nigh prone to exaggerate the present sphere of man's individual responsibility. We are, like many Christian ministers, almost disposed to accuse mankind of intentional wickedness, and to say: "If men would but will better, they would do better;" and because they do not so will, we are ever and anon tempted to add, "Men should be compelled to righteousness," for they are radically capable of lives vastly more noble and harmonious.

This view would lead to censoriousness and vituperative denunciation. It would put us out of tune with that "charity which thinketh no evil." It would inspire us with pity, sarcasm, irony, hatred, and contempt. It would embitter all our love for mankind. It would cause us to quote condemnatory sentences from poetry, old sermons, and the Bible. But we are saved from all this misfortune by the discovery that man's character is not self-made, but that it is a reflection of a combination of causes, over which, for the present, he has little or no control.

The truest Spiritualism is beautifully practical as well as gorgeously theoretical. The truth is known by its adaptations. With regard to man the truth is, that, while he is interiorly organized for complete self-government and unlimited moral responsibility, he is not, in the present state of his spiritual development, much more than a creature of external interests and circumstances. His outside, every-day, working Character is a product of his most positive surroundings. He acts and manifests traits in accordance with the circumstances by which he is compelled to exist. He is a creature of motives, of interests, of downright physical necessities. "A hungry man knows no law." It is a fixed maxim of common sense that "evil communications corrupt good manners." Good manners crop out from good morals. Morals, therefore, are first injured and corrupted by vicious associations. May it not be equally true that *discordant external circumstances will unbalance the most harmonious character?*

The fertile sources of human discord lie back of man's birth; also in the sphere of immediate social and physical circumstances. Christians accuse the human heart of evil. We do not. Because the world's evil and iniquity are traceable to the world's outward constitution. It is vain for moralists to expect spiritual beauty from persons overburdened with grievous wrongs and misfortunes. In vain may preachers enforce the practice of great principles upon minds well nigh crushed by the weight of poverty and injustice. Thousands all around us are compelled by their circumstances to expend their noblest energies for the preservation of mere physical life. Others are tied to the wheels of perpetual servitude. And yet others there are, who, by the accident of birth, or from some other equally external cause, are bound day and night by the tyrannical will of heartless and soulless men—themselves the inheritors of wrongs and injustice done to ancestors long since departed. With body cramped and diseased, with soul fettered and stricken down from its natal hour, with every external condition unfavorable to the development of the gentle virtues and beautiful attributes, what can you expect? What may we hope from delicate women who are wasting vital forces by the incessant effort to maintain themselves and families? What shall we expect from finely-constituted men, who, being crushed by poverty, are straining every nerve to provide for the bodily necessities of those dependent upon them? Beautiful minded persons there are, who have not *an hour* to devote to mental improvement. They are chained to the labor-wheel of servitude, and compelled to toil year after year for the benefit of others.

We wonder not that there are vicious men and down-trodden women. We should wonder exceedingly if men and women were not just what they are: the reflections at once of their inherited organizations and social circumstances. The higher spiritual spheres look down with unbounded charity upon all human kind. Higher intelligences see that "the just and the unjust" are alike the effects of antecedent and existing causes.

That the world cannot be reformed by merely appealing to the moral affections, is clearly demonstrated. The well-known and successful "Five Points Mission," of this city, was prosperous only when the *physical conditions* of the abandoned population were examined into and absolutely improved. Just in proportion as the bodily circumstances of the depraved were reformed, in that same proportion did the inhabitants become contented and virtuous. All preaching of Christian morals, prior to the physical improvements of the down-trodden, was like water spilt upon the sand which yields neither fruit nor flowers. In this manner it is demonstrated that physical slavery is the cause of moral deadness and degradation. The soul is bound with the body; they cannot, in this life, be separated.

Now, if you apply this reasoning to the political institutions of a country, what will be your conclusions? You will at once decide that great national progress is impossible under political injustice and tyranny. A corrupt government is better than no government at all, because the people, under its diabolical oppressions and obvious wrongs, make progress by their efforts to overthrow its foundations. But such progress is replete with cruelties and excesses. The people shed their burdens by violent efforts. They break their chains by destroying those who fettered them. All this is sad and lamentable, but it is natural under certain physical conditions, and every despotic government will know it in the coming future.

But there is a more excellent way "to overcome evil." A human soul gains nothing by *fighting* the conditions of evil. If you quarrel with evil, it will overcome and vanquish you. It will compel you to become like itself—ugly, bitter, hateful, sarcastic, ironical, combative, cruel, malicious, murderous. However good in the start, in the end you will be like that which you have resisted and fought. Suppose a traveler should lose his path in the mazy wilds of a vast forest. He cannot find his way out, and a lonely death in the wilderness seems the impending fate. What would you think of him were he to commence cutting down the trees as the best means of escape? Would he not die in the useless labor? If he should spend the same number of hours, and the same amount of strength, in an intelligent effort to discover his way back to civilization, would you not approve and applaud the wisdom of his undertaking?

So of every other human mistake, misfortune, or evil. Do not spend an hour in fault-finding and combativeness, but go straight over the "evil" to that condition which is "good." Do not fight disease, but fix your eyes upon the conditions of Health, and put your body in a situation to be improved and strengthened. Do not fight a private vice. All efforts to kill the vice will result in your defeat. You may "resolve and re-resolve," but you will "die the same." And why? Because you have not attempted to overcome (or to go over) evil with good, but instead, have spent your noblest energies and sweetest hours in vain attempts to conquer evil with evil. If your ship is sinking, take to the life-boat, and pull for the shore. If your neighbor be unjust to you, do not give the bad example of fighting with him, but straightway put yourself beyond the possibility of a like invasion. It is stronger and far more progressive to do right than wrong. It is far easier to contend with an evil than to inaugurate a good; and most people take the easiest, and consequently the surest, road to failure. A conquest is rare. If a man have an error, what will you do for him? Will you quarrel with him and with his error, or will you present a new truth to his mind? Suppose the case is your own, and you have an error or a vice. (You have many, doubtless.) What will be your wisest and straightest rule of conduct with reference to yourself? The wisest, we affirm, is to progress away from it and them. Suppose you have the ungentlemanly habit of using profane language. Now how will you break up the habit? By endeavoring to remember what words you should *not* use in conversation? Far otherwise. The true way is to make up your mind as to the words you *will* employ in communicating your thoughts. Suppose you are an imbiber of strong drink, and you wish to cease the use of it from this hour. (We hope you do so desire.) What is your best course? To fight against the propensity to intoxication? No, Brother. The only certain way is to turn your back upon

10*

yourself, and decide, not what you will *not* drink, but what you *will* drink, or die in the attempt! This is overcoming evil with good. All progress is positive and affirmative; all failure is negative and conservative. The first is characterized by the development of superior conditions; the latter by an attempted overthrow of conditions that are low and evil; and the result is, that while the latter is struggling to overcome evil with evil, the former is really *transforming evil into good.*

CHAPTER XIII.
HOW TO DO GOOD.

We receive questions substantially as follows: "How shall a believer in the Harmonial Philosophy most advantageously live for the world's permanent progress?" Or, "How shall a true friend of Progress devote his energies so as to accomplish the greatest amount of good to his race?"

Glorious questions! A truthful answer is demanded by every noble aspiration, and we shall, with all possible brevity, attempt to evolve a few thoughts, which, we trust, will find lodgment in the vast storehouse of intelligence.

Man, at last, stands upon the threshold of a true civilization. We mean exactly what we say—"threshold." For the best one among us has not entered the temple further than the middle of the spacious vestibule. Behind him is the Cauldron of the Past. Past opinions, thoughts of olden times, antiquated religions, theories of departed philosophers, fables of the ancient ages—all, in one conglomerated mass, is boiling, seething, bubbling, fermenting, over the fire of purification, in the Cauldron of the Past. The fire of emancipated Reason is distilling the clear wine of Truth from the fruit and grains of departed generations. Thousands in the march of civilization retain tenderest sympathies for the traditions and doctrines of the earlier periods. They shout with pain at the sight of decomposing creeds and falling temples of sacred error. Those who stand on the door-step of spiritual civilization, beholding the destruction of the useless *forms* of past revelations, are the first to cry out against those just before them in the onward march. But those who have penetrated the vestibule of the Future era on earth, are the first to behold the principles, and consequent progress, of the human mind. To such, the vast realities of harmonial brotherhood, crowned with the diadem of distributive justice and universal liberty, loom up in the immediate future of civilization, like the temple of happiness in the heart of Deity.

The best man, the wisest woman, the boldest thinker, the most lucid poet, the sweetest philanthropist—none of these are wholly civilized. To send our institutions into Africa, to project our systems of religion into the Japanese Empire, to fix the manacles of our incomplete government upon less powerful nationalities, to hold up our examples as the best for the heathen, is to extol a green civilization as though it were adapted for common digestion. It would be not less injurious to other peoples, than unripe plums to the stomachs of children Flushed by the rapid development of successful machinery overcome by the intoxicating stimulus of commercial expansion, made vain by the discovery of unlimited powers of thought, the vanguard shout "Victory!" and "Christian Civilization!" when, in sober reality, we are but just stepping out of barbarism into the vestibule of that "good time," which shall be a grand joy unto all people.

We live in a glorious age, not because this is a civilized epoch, when liberty and happiness are appreciated and enjoyed equally by all, but this age is glorious because it is *the bud*, just before the era of flowering into magnificent beauty. In their highest efforts of thought, men have not advanced much beyond a weak juvenescence, a sort of spiritual infancy, and much progress is yet required in order to crown them with the dignity and glory of adolescent wisdom. Therefore, to the true friend of man, all life is practical. The real, true, harmonial soul, does not speculate on the nature of his aspirations, to the neglect of his co-ordinate duties. He wants to know what he can advantageously do to hasten the Better Day, and to fit him individually to unfold, without pain or regret, in another sphere of existence. He believes in the principle of progress. He understands that it will force him onward, even though he should not attempt to harmonize with it, just as the ocean tide wafts forward the drift rubbish upon its bosom; but he naturally scorns the imputation of being only a *block*-head, or a man of rotten *wood*, and therefore asks: "*What can I do to accomplish the most good for the Race?*"

Noble Reader! Your soul hath honored its sublime constitution in the putting of a question so benevolent and angel-like. We love thee now, more than ever, with a growing love, full of joy and fraternal peace. We come close to thee, very near to the inner life of thy being, and whisper a few words of counsel. Dost thou hear them?

Be not deceived. Thou art in the march of Civilization, but *not civilized;* a believer in Spiritualism, but *not spiritualized;* a worshiper of Truth, but *not truthful;* a lover of Wisdom, but *not wise;* a seeker after Happiness, but *not happy;* a pilgrim in the ways of Progress, but *not progressive;* and so the sublime picture of life is broken in fragments at thy feet; the whole of creation is often less stupendous than a small part before thine eyes; the good God of Nature is sometimes hardly equal to a man; and, in the very nature of things, thou art just as far from the Harmonial state as a child is removed from the condition of Manhood. What is to be done?

Answer: BE A MAN. That is, whether brother or sister, BE A WHOLE SOUL. How? To all the faculties of thy being, BE TRUE. Be true to all the dictates of thy superior powers. Speak the truth; do not falsify. Get knowledge; do not propagate ignorance. Be spiritual; not merely a Spiritualist. Exemplify freedom; do not be a slave. Become civilized; do not remain in barbarism. Be happy as possible just now, in your circumstances; do not put off the hour of happiness. Make progress; do not merely preach it.

'Tis easy to counsel thee, dear reader, but how hard to practice! Nay, quite otherwise. Truth-speaking is as easy as truth-writing. It is more easy to do right than wrong. Motion, to thee, is easier than stillness, even as happiness is more pleasurable than discord. Doubtless you will fail a few times at first, as children frequently fall between the cradle and independent walking, but invariably the "right comes uppermost" after successive efforts to bring it.

Go forward and upward! The heavens are populated with spirit-hosts adequate to thy necessities. Make the march of mind a sacred reality, not the dream of the over-heated enthusiast. Do your work nobly, with a spontaneous love for it, and with energy. Do not look timidly back upon the boiling cauldron of error. The fire of dispassionate Reason will purify the hell of all past deeds. It will burn and destroy your vices, and all crimes, but not until you assist in kindling the fire. Every soul is required to place some fuel under the distilling crucible. Another man's merit will not make you happy. It is ordered, and wisely, too, that no idle man may long enjoy the food and fire of the industrious. Bring fuel to the fire of reform, therefore, and work to burn up your own evils. Set your alcohol on fire. Destroy all your noxious weeds of vice. Let the furnace of private redemption burn hotter and hotter, until every personal discord is consumed. Fix your whole heart firmly upon what your higher faculties admire, and do their bidding.

CHAPTER XIV.
MIND AND MATTER.

EARTH's powers and principalities exclude most men from the society of poetry and eternal principles. Matter is a powerful and controlling God; it is the "prince of darkness" to millions of our throbbing humanity. Matter clings and clusters heavily about man's interior Life; it is the dead freight of his perilous voyage from the cradle to the coffin. Men are necessitated to worship at the shrine of Matter. They make it the chief object both of masterly effort and spiritual contemplation. Thousands reverence Matter incessantly. They bow down before its altars. They bring to it many offerings—tithes of mint, anise, cummin, and lip-service—covering its temples with everything within the power of man to bestow; with scientific art and the works of genius, with developments of the noblest talents—with everything, even life itself.

Mammon is but the servant of matter; matter is but the servant of soul; soul is but the servant of spirit; but, in this lower world, it happens that spirit, and soul, and matter, are the servants of Mammon. No human soul is independent of its material surroundings. All human "Life is real" bondage to matter. Individual "life is earnest" in overcoming this bondage. But "the grave is not its goal," because the soul is not destroyed by its environments. The physical circumstances of the spirit are negative at last; but they are absolute and positive in this sphere. Matter is the mind's jailer. Want is the overseer who lashes the prisoner into his daily labors. 'Tis the mandate of matter which the mind obeys nine-tenths of earthly time. The sight of objects, the taste of flavors, the smell of odors, the cognition of sensations, the hearing of sounds—thus the spirit looks out and lives through the *grated windows* of its prison castle. A defect in either sense is so much subtracted from the liberty and capacity of the mind. Deficiency in blood or brain, and misplacements of either material, or the slightest excess in any department, are recorded mathematically upon the ledger-pages of the life book.

The universe, with its beauties, and laws, and harmonies, is *nothing* to the idiot mind caged in matter. The gorgeous heavens, with their unnumbered systems of suns and stars, are *nothing* to a soul bowed down by the daily drag of material necessities. The ponderous globes of space, so attractive to the uplifted mind of the philosopher, are *nothing* to my brother who makes a God of gain. Matter and money surround him on either side. He drives through his surroundings, and then they drive through him; and so goes his daily life, "to the last syllable of recorded time."

The fair sky of heavenly truth never covers the earthly mind. Angels do not dwell in the shades of pandemonium. Matter is the raw material of Heaven and of angels. Strange paradoxes! The world of matter is the region of discord. The myriad forms of evil originate in the realms of Matter. The history of our beginning is a salutary history, because it teaches the lessons of progression and imperfection—how chaos precedes order; Matter, mind.

But it is a pleasant thing to die! Why? Because the countless shades of matter, like storm clouds and dreams of prison life, begin to move off and forever away. Matter, the soul's prison, is abandoned. The spirit in quiet looks upon the dim substance stretched and cold on the earth below. The dark broad mountains of matter, where the thunders of earthly discords rave both day and night, are forsaken or exchanged for flower-clad hills, "eternal in the heavens."

It is a pleasant thing to die, and to join the peaceful brotherhood of the upper realm. The Divine Mind, whose infinite powers and principles fill all the temple of immensity, is seen by spirit. Matter is incapable of contemplation; yet it is the deep-hewn valley in which soul is cradled. The soul is the chariot of the golden spirit; but alas! in this world, Matter is both the driver and the steeds. Matter is molded into shapes replete with grandeur and sublimity; but the power to cognize and enjoy is inseparable from spirit.

It is a pleasant thing to die, because, by a natural going forth of the spirit, at the appropriate period of its history, the evils of matter are more readily comprehended and overcome. The music of spiritual waters floats into the new-born soul. The sickening shadows of terrestrial ignorance and misunderstanding depart among the broken urns, behind the curtains of time. The principalities of falsehood lose their power. They fade away. The pure light of a measureless firmament shines down into the reasoning faculties. Whirling globes, supporting innumerable forms of life and beauty, fill the immensity with the glory of God. Harmonies of the affections, touched by the awakening love of celestial fingers, come up and down like the breathings of truth, causing the immortal hills to sing like birds of a thousand voices. Outspreading landscapes become vocal with an abundant harvest of eternal love-lessons, too pure for earthly language to embody. And thus, the unearthed soul is sent to school among the angels of truth and the Titans of wisdom.

It is a pleasant thing to die, when the death is natural, because the soul "makes a Sabbath day's journey" toward Deity—gets nearer to the central Fountain of everlasting life—nearer in the sense of realizing more love and acquiring a higher knowledge concerning the spiritual laws of the universe. The kindling fires of infinite life light up the trans-mundane pilgrim with a larger and diviner comprehension. The great cycles of the world's progression appear like changes in the performance of an operatic drama. The rise and fall of empires seem not more important than the shifting of scenes in a theater—the lights and shades of an immortal picture.

It is a pleasant thing to die, and to get out of the prison of engrossing and heavy matter, because its *chemical* transactions emancipate the spirit from the imperative besetments of hateful appetites. Although the soul retains the effects, sad and many, arising from the multifarious transgressions of the principles that are indispensable to its progression; yet, by the fact of chemically altering the relations subsisting between soul and body, the spirit is measurably empowered and inspired to rise above its ruling earthly passions. With the body goes tobacco, alcohol, stimulants, &c.; and with death comes the power to be larger and happier. Some minds are vicious because of physiological defects. Brains sometimes are imprisoned by a malformed skull, and spirit is embarrassed by a hampered brain, and *character* is deformed as a consequence; so that, in contemplating our common humanity, it is wisest to put down a large amount of evil to externals, which, in this life, are positive in begetting personal manifestations. An accident has been known to jar into life certain portions of a long-slumbering brain, whereby the prosy person was at once converted into a poetic genius and partial musician. Imperceptible alterations in the cerebral polarities will be followed by special changes in the character and habits of the individual. Sorrowful persons may suddenly become joyous and gay; drunkards change into the finest examples of temperance; vulgar souls turn into the paths of refinement; and thus, "in the twinkling of an eye," it is possible for Death to elevate the character and multiply the opportunities of a man. Even here, under the magic touch of human magnetism, the ignorant soul is *suddenly* converted into the embodiment of surpassing intelligence; and by means of the same transforming influence, the mouth of the dumb is opened, and the slow tongue is made to move with the lightning flashes of eloquence. If a few passes of the human hand can work changes so instantaneous and so marvellous upon a human soul, while yet in the body, what are we not authorized to expect when Death bursts the "prison-house of clay," and gives the mental powers liberty to run to and fro "through the halls of creation," in the natural exercise of all constitutional rights and inclinations?

Yesterday we climbed to the loftiest summit of a dark, broad, and beautiful mountain. We sought a solitary dwelling place beneath the shadow of many trees. The beetling cliffs lifted their stately summits on either side. The music of the deep valleys below filled the temple with sacred melody. The far-off silvery clouds, floating between our upturned eyes and the summer sun, seemed to welcome our thoughts to the worlds on high. We there obtained a wondrous vision of truth, and law, and soul, and matter; and, for the thousandth time, we acquired a lesson from Mother Nature to this effect—"it is a pleasant thing to die."

PART II.
PHARMACOPŒIA.

DIAGNOSES AND PRESCRIPTIONS.

The leading pathological propositions set forth in this volume, by which we trust all progressive physicians and patients will be guided in the application of our various prescriptions, are :

1. That all diseases originate in a disturbance of the soul-principle, which consists of Motion, Life, and Sensation.
2. That the effect of such disturbance is a development of *local suffering*, invariably in the region of the greatest *previous* weakness, or where there is the most constitutional tendency to disease.
3. That the concomitants are an increase of *motion* and of *temperature*, or else a reduction of temperature and a diminution of motion, in the fluids and forces of the parts assailed.

From the above propositions we are authorized to conclude that all diseases are characterized by *active* and *passive* symptoms or states ; and we hereby suggest, with all due deference, that friendly physicians and all patients be regulated by these principles in the administration of appropriate remedies.

Spring Time Diseases.

When we emerge from the Winter months into the terrestrial thawings and atmospheric changefulness of the coming Spring, the eccentric action of this peculiar season upon the fluids and solids of the body, is irregular and disease-generating. Dense fluids descend from the brain and lungs, and disperse through the lower viscera—laying the foundation for various stomach, membranous, and liver disorders, resulting in obstructions, diarrhœa, congestions, and inflammations. Meanwhile the lighter and rarer fluids ascend from the bowels and liver, and diffuse themselves through the solids and nerves of the throat, face, and brain. These currental and vitalic changes occur in every human body, just as surely as the upward flow of sap in trees, at the very beginning of the Spring months. In the perilous passage between Winter and Spring more children get sick, and more adult invalids depart for the next Sphere before they should, than during any other season of the year.

PREVENTION AND REMEDIES: May holy angels throw their strong, white arms around the multitudinous little ones of earth; and may the understandings of all parents be opened to the true ways of life and health. Especially do we pray for the protection and conservation of the diseased, the suffering, the famishing, and the unhappy. Let them organize themselves into systematic, thorough-going, wide-awake defenses against the approaches and invasions of the ruffian, Gen. De Bility. Each human being is provided by the Father and Mother with ample constitutional means of resistance. Whole troops of WILL-POWERS are garrisoned in every visceral organ of the body; indeed, it is literally true to say that every organ—from the top of the head to the depth of the abdomen—is naturally an impregnable fortification.

Incredible as it may seem, it is nevertheless true, that man's body is a strong tower of defense, a fort of marvellous construction, which no atmospheric changes can affect or touch, if the proprietor but understood the shielding power and vast sweep of the invisible Will. Faith, knowledge, imagination, affection, intuition, and fidelity, enter as ingredients into the composition of WILL. It is not what phrenologists term Firmness, it is not an obstinate and dogged principle of stubbornness in the mental constitution—not a "will" and a "won't" propensity—quite otherwise; by Will we mean the concentrated whole-hearted-ness of the brain and soul-life flowing like sunlight within the blood to any desired part of the physical economy, distributing the fertilizations of sublime health and strength through every crevice, and tissue, and nerve of the dependent frame.

We admonish each of all our readers to put forth this regenerating, this anti-suicidal, this immortal energy of the spirit. Suppose that Spring is about to burst upon you, with its varied terrestrial and atmospheric characteristics; with its fickleness of sunlight, hesitancy of temperatures, and with its changefulness of electrical and magnetic currents. Go forth, then, armed with well-balanced manhood! Like a strong, well-formed, beautiful woman, go forth lovingly to greet and cheer on the Spring; do not remain in the house all day, whether sick or well, but walk forth, panoplied with the WILL-POWER, and thank the universal God of Nature that in him you "live, and move, and have a being." Be always very thankful; let your face shine with gladness, your cheeks blush with youthful vigor, (although the record may be that you are more than sixty years old,) and the host of overseeing intelligences will drop a "new lease upon life" in your heart, which will then steadily beat against your bosom as the nearest friend you have. Do not get a "bad cold;" do not permit yourself to cough; do not get the habit of taking on rheumatic pains. All these ailments may be *prevented*. Bathe your feet in cold water before walking; keep them protected against the dampness of the ground; breathe deeply while walking, allowing the breath to escape only through the nostrils; swing your arms, firmly shutting and opening your hands occasionally; be strong and energetic, not flexible as India rubber, in putting forth muscular effort; get honestly warm and generally fatigued—earn and deserve your weariness—then, on returning to thy habitation, go into thy chamber, lock the door against every external intruder, and resign thyself to the nourishment of slumber. Nature will faithfully awaken you at the right moment; then, whether sleepy and rested or not, arise and resume the business of the hour.

During Spring months it is better to eat almost no meat. Whether sick or well, this counsel is applicable to you. Eat various kinds of simple puddings for dinner, only *one compound*, with bread and butter, at any meal. Farmers and merchants, mothers and children—each and all, better keep "Lent" and grow healthy by fasting in the early Spring months. The following is a good dinner for two days of each week: "Take half a pound of bread crumbs, half a pound of potatoes, boiled, quarter pound of suet, chopped fine, two eggs, well beaten. Mix with milk, and boil three hours." A large family, of course, would require a larger pudding than these proportions indicate.

Abolish the demon, "coffee." Do not spend another copper to obtain this copper-colored *enemy* of lung, liver, stomach, bowels, throat, brain, and reproduction. Nature will allow you to use black tea, not too strong; never oftener than twice a day, seldom at night. Let all families manufacture a beverage for Spring drinking; to be used at any time, even at meals, instead of warm drinks, ale, or porter; for it will give a healthy fluidity to the blood before warm weather, open the bowels moderately, and assist the feeblest digestion.

A Spring Beverage.—Eight ounces of sarsaparilla, three ounces of liquorice, six ounces of wild cherry bark, half ounce of mandrake, one ounce of gentian, and half a teaspoonful of each cinnamon and red pepper. Boil in three gallons of rain water until the quantity is about half reduced. Let children and adults drink a wine-glass of this whenever thirsty. Do not sweeten it much; nor allow yourself or little ones to indulge in sweets.

Better eat an orange before breakfast than at any other time during the day. Give your little children oranges without the pulp or seeds. They cannot easily eat too many oranges in the Spring. They contain very rare properties of strength to the sick and debilitated. Dyspeptics would do well to walk before breakfast, and eat a couple of oranges, breathing as directed in the meantime. Consumptive and bilious persons may obtain much relief from the free use of oranges anywhere between early rising and the second meal. Let every reader of the New Philosophy give evidence of the glorious faith within him.

Origin of Skin Diseases.

Our philosophy of diseases of the skin differs, in several essential particulars, from the received theories of high medical authorities. It is deemed necessary to explain our impressions concerning these distressing afflictions, in order to answer, at one and the same time, scores of letters from persons *sorely* afflicted with external diseases.

The opinion most generally received is that all eruptive affections originate from unhealthy conditions of the blood. Pimples, pustules, blotches, scabs, itch, rash, salt rheum, measles, scarlatina, fever sores, scrofula, cancers, tumors, erysipelas, small pox, &c., are attributed to depraved and poisoned conditions of the blood. Thousands of patent nostrums are manufactured expressly to take pecuniary advantage of this prevailing error and consequent weakness among men respecting the blood-origin of all skin diseases. There is no exaggeration in the statement that, notwithstanding the barrels of empirical syrups and quack tinctures made and swallowed per annum, in order to cleanse and purify the blood, the number, and variety, and intensity of skin diseases, are constantly and alarmingly on the increase. All through civilized society we observe the palpable evidences of scrofula, erysipelas, cancer, small pox, &c., and the victims, as a general thing, receive but little assistance from the leaders of medical science. Manufacturers and venders of infallible medicines require their patients to take their medicine according to printed directions on the label. These printed directions, being literally translated from the mysterious depths of the internal sense, in which their authors wrote them, would read thus: "The manufacturer and proprietor of this Sovereign Remedy (for all skin diseases) cannot guarantee a permanent cure unless the directions be faithfully and strictly followed; and these important and authoritative directions are, that this infallible preparation be shaken and taken '*internally and eternally*,' on which conditions alone the proprietor hereby promises to refund the price of the bottle if a perfect cure be not effected."

The primary origin of skin diseases is the disturbed condition of the vital principle within the blood, by the positive and negative operations of which, the bodily fluid is kept warm and ceaselessly in motion. This position is established by the fact, everywhere well known and acknowledged, that all persons are not susceptible to the same skin diseases. Certain temperaments do not absorb small pox, the itch, salt rheum, &c.; while others will take on these and yet other skin diseases at the slightest exposure. The depravity, or derangement, originates first in the life principle; then, as a logical sequence, the external will receive the form and embodiment. The soul (which is the vital covering of the immortal spirit,) is primarily disturbed; and other consequences will follow, "as the night the day," until the *effect* itself becomes a disease, and subsequently the *cause* of like disturbances within the soul. The blood is a dependent *subject* of the controlling, animating principle; and the diseases of the body, therefore, are indications of diseases existing first in the soul.

When a disease ultimates itself, and begins to organize a life of its own, it is then self-evident that the truest treatment is to disorganize the new comer and at once dispel it, somewhat as you would drive a wild beast from the threshold of your house. But in the first instance, no treatment would prove efficacious, unless directed toward the vital forces of the entire organism. The true medicines are food, air, light, exercise, sleeping, &c., all which come within the jurisdiction of love, WILL, and wisdom.

But suppose from a small bruise or scratch on the foot, shinbone, knee, or finger, a sore is organized and established. Do you not see that the state of the blood is not the cause of the ulceration? The blood may be perfectly healthy, and yet an injury done to the anatomy of the skin may be, and often is, followed by the organization of a cancerous body, or of inflammatory and malignant sores. The anatomy and physiology of the skin are sufficient to demonstrate that, irrespective of the condition of the blood, the injured structure of the cuticle will organize a disease of its own.

It is plain that, in many cases of eruptive diseases, the directest treatment is the wisest. Cancers frequently return, because, although sometimes perfectly removed by external applications, the membranes and blood-*vessels* (not the blood) of the cuticle are not perfectly restored to their original state. The deranged and mutilated parts, after the tumor or cancer is removed, commence forthwith to malconstruct the bodily nutriment into another nucleus, which, in time, is likely to become the center of a similar formation.

Your blood might be as healthy and pure as that of an angel, and yet, if the membranous layers, the delicate nerves, the refined tissues, and the conducting vessels of the skin be disorganized, either by sickness or accident, it will be next to impossible for you to avoid the organization of some external disease. The only possible preventive is the judicious use of your Will, and the restorative magnetism of a magnetic hand. Sometimes the act of *rubbing* a chronic sore with an emollient salve, or simply by bathing and dressing it with the gentle hand, imparts a healing power to the parts, which the maker of the salve appropriates and attributes to the virtue of his prescription. The cause of the restoration is human magnetism, and the Will. The arteries and veins form a net-work on the true skin; it is also supplied with lymphatic vessels, with glands, and capillary nerves; so that, in cases of accident or injury done to the delicate parts, the risk is very great as against the welfare of the system. The skin is full of glandular lungs, so to speak; its health depends constantly on natural ventilation. If the exhalation of bodily vapors be retarded by lack of cleanliness, or if the inflowing magnetism of immensity be excluded by a like cause, the consequences are direful on the brain, and, ultimately, within the soul-forces of the higher organism.

We make these remarks as practical hints to all patients who apply our prescriptions for cutaneous diseases. We wish the philosophy of skin affections to be practically comprehended —namely: that the *condition of the skin* is of more importance than the state of the blood—and that we do not propose to purify the blood by syrups, but rather to balance the vital forces and restore the structure of the skin to its original condition.

Acute and Chronic Cases.

"It is a fearful thing to teach," says that remarkable book entitled the "Healing of the Nations," through the inspired mediumship of Mr. Linton. "The secrets of true knowledge are hard to find, and when found are hard to be explained. Hard to find, because they tend, step by step, towards the Center—God; and hard to be explained, because all things are as rays of Him—and He cannot by aught below Him be comprehended."

Not less difficult is the discovery and explanation of the true causes of disease and bodily misery. Here, for example, is a patient, suffering with what several able physicians, in general phraseology, term "a disordered liver." One authority says there is in this case an excess of bile, a torpid state of the liver, and mucous derangement of the stomach; another asserts an excessive activity of the liver, a deficiency of bile, and sympathetic disease of the digestive system.

SYMPTOMS. Yellowness of the white of the eye, dry skin, bitter tongue, thick saliva, dull headache, redness of the nose and chin, occasional flushings and nausea, nervous irritability, restlessness, deficient evacuations from the bowels, and irregular appetite for even the most agreeable articles of food. These symptoms appear and disappear, rise and fall, with considerable periodicity of movement. About every twelve days they put forth premonitory indications. They increase in number and violence, and culminate with pains in the back, shoulders, &c., and then subside, all within four days from the commencement. But the patient is never in an amiable mood. Is it chronic inflammation of the liver? His mental irritability is incessant and extreme. Sometimes he fancies that his heart is badly diseased. At one time he is afflicted with impatience and disgust; is unreasonably peevish and flagrant in his combative suggestions; anon, without any apparent cause, he is mercurial and impetuous in disposition; and then comes on a period of gloominess and bitter depression of feeling, enough to repel the tender watchfulness of a sainted soul.

THE REAL DISEASE.—What is the real secret of all these signs and symptoms? We have examined many such cases, and therefore know that it is not a disease of the liver, *but a chronic inflammation of the duodenum*. This part is the short channel between the stomach and the small intestinal system. The gall ducts empty the bile into the chyme just at this important junction, and right at this point, also, the ganglionic magnetic energy is dissipated, resulting in muscous and nervous derangement of the liver and bowels. This condition is very common among American women. The great ganglionic net-work, of magnetic and electrical conductors, is incidentally disordered and incapacitated. All the biliary signs and symptoms are traceable to this diseased state of the *duodenum;* and the mental irrascibility and quarrelsomeness are inevitable where *duodenitis*, in chronic form, exists and prevails.

THE CURE.—Shall such a patient experiment with the pneumogastric forces of his own system? Certainly, but it must be accompanied with obedience to the laws of Health in other particulars. By stimulating food, passion, and excesses of various kind, it is easy to overload the liver with blood. It may secrete too much bile for a long time; then, again, it is hardly able to generate sufficient for its own support. The chyme is neglected. The consequence is a redundant action in the duodenum—making the feebleness and oppression of digestion worse and worse.

What is the treatment? Besides using the Self-Healing principles, the patient must abstain from fluids for breakfast, and omit his supper, taking only a roasted potato, with a little bread, not later than five o'clock, P. M.; the dinner may be composed of well-known healthy substances; no drinking while eating, and at no time in quantities sufficient to distend the abdomen. Two hours after this meal he should put a large cold-water bandage about the waist, well enveloped by a dry woolen cloth; then attempt to get a little sleep; after which take a walk, or some other sort of physical exercise. Pursue this method day after day, until the inflammation is reduced in the duodenum, but avail yourself of human magnetism frequently. It is highly necessary for this class of patients to abstain from all hot-water bathing; but a sweat by vapor, perhaps thrice a month, is not unprofitable in disorders of the liver and blood. And now, gloomy sufferer! cheer up and act like a man. To thee the poet hath well said,

> "Many a foe is a friend in disguise,
> Many a trouble a blessing most true,
> Helping the heart to be happy and wise,
> With love ever precious, and joys ever new!
> Stand in the van,
> Strive like a man!
> This is the bravest and cleverest plan,
> Trusting in God while you do what you can.
> Cheerily, cheerily, then! cheer up."

Your frettings and quarrelsomeness are signs of discords within you. You profanely imagine that the Almighty is concerned in the apportionment of your sufferings. Away with such conceptions. Be stoutly honest henceforth, and say: "I have sinned against the laws of health, both knowingly and innocently, and the true consequences are upon me." Shakspeare, in the play of King Lear, says: "This is the excellent foppery of the world, that when we are sick, (often the surfeit of our own behavior,) we make guilty of our disasters the sun, the moon, and the stars;—as if we were villains by necessity—fools by Heavenly compulsion—and all that we are evil in, by a Divine thrusting on."

How to Exert the Will.

QUESTION.—"*Dear Sir:* You teach that every disease may be cured by the action of the *Will*, and more particularly 'general debility.' .... I can very readily understand the action of the 'Will' in the common sense of the term; that is, to *will* anything done is to do it. For instance: I have a very severe pain or rheumatism in one of my members. I may wish and desire it well, but that won't do it. But if I procure the *proper* remedies, the cure is accomplished. Now is it not done by *acting out* the *Will?* Would it not be like the wagoner when he stuck fast in the mire? He called upon Jove to help him out. But a voice called to him and said: 'Apply thy shoulder to the wheel and whip thy horses well, then call upon Jove and he will help thee out.'"

ANSWER.—We are persuaded that our interrogator cannot easily misconstrue the significance of the terms "will" and "willing," as applied to the action of mind in the treatment and eradication of disease. WILL is the diplomatized plenipotentiary of all the impulses of love, and it is equally the agent of all the intelligent mandates and wishes of Wisdom; so that when the whole mind, with righteous intent and totality of purpose, directs its *nervauric force* upon any part of the body, the effect cannot but be absolute.

Blood constantly goes in large quantities to the brain, in order to give stamina and energy to its constitution and temperament. If the front brain be principally used, then the blood, (or rather the magnetism thereof,) will concentrate at the head, leaving the extremities and visceral organs comparatively destitute of their appropriate proportions of the vital power. On the other hand, a constant employment of the muscular system of the extremities, including the vital organs, will attract the magnetic forces from the brain, and, as a consequence, continuous vigorous thinking becomes well nigh impossible.

Now we find by long and patient experimentation, that the WILL, under the direction of both *desire* and *judgment*, is capable of throwing a magnetic force upon any muscle or organ within the body. The true way to begin, is: to act upon the *voluntary* muscles of the extremities while lying on your back. Cause the toes and fingers to open and shut simultaneously. In a short time, both extremities will become warm and magnetic. Then swell, and stiffen, and relax, alternately, the muscles of the legs and arms. Deep, long breathing is indispensable to success. At the next experiment, fix your will upon the longitudinal muscles of the abdomen. Cause them to dilate and fall steadily and firmly.

After a few trials you can affect the liver, stomach, and heart, just as surely as you open and shut your hand. Do you not know that man's Will-energy is terrible in power? In the medical realm the Will's *self-healing* attributes are little known. We know that, with many good people, the theory of exerting Will upon their sick bodies is an impracticability. But even so it was for a long time wholly impossible for Dr. Winship to raise 1,000 pounds from the earth, but he has done it again and again. Time has been when a tenth part of that weight would have fatigued him. Do you not recall the expressive anecdote? "How much do you weigh, Jonathan?" asked a frail-bodied merchant of a young New England farmer. "Wal," he replied, "commonly speaking, I weigh one hundred and forty-five pounds—but," he added, stretching his fine form up to his full hight, and dilating every muscle with the magnetic energy of Will—"*but when I'm mad I weigh a ton!*" Does not our reader weigh a ton, also, and that, too, in a good cause—in the direction of Health, for instance?

Automatic Forces.

In examining the mind's internal mechanism, we get at not only the action of organs, but also discern the nature of the action. Each part of the mind diffuses a particular influence all over the constitution, and the influences that have emanated from all the parts constitute "sensation," or the lightning of the nervous system; and, inasmuch as human beings are organized upon the same principle, so it happens that an influence imparted to another awakens in that other effects analogous to those felt by the one who imparted it. Thus a combative person, on imparting his organal influence, will cause another to experience identical sensations. The same is true of every organ.

These facts are familiar to modern psychologists. They stand in the gateway between heaven and earth—preventing at once too great credulity and too much doubt—for such facts demonstrate, not only that the nature of man is double, but also that he is not *the cause* of all spirit phenomena. The automatic hemisphere of mind is quite as marvellous as is the counter-hemisphere of voluntary powers; and, when truly studied, man becomes as much of a wonder before death, as when he returns in the estate of spirit.

It would seem that man's own spirit ofttimes continues the process originally instigated by invisible intelligences. They may diffuse an influence upon his nervous system, which, entering into chemical combination with the sensitive elements, they (the spirits) can neither control nor extract from their subject. It is evident that many spirits have little knowledge of their own abilities to control the influence they cast upon their fellow men. The consequence is, that what should be voluntary, and under the control of the wishes or will of the subject, becomes, instead, *automatic* and beyond management.

Man's Telegraphic Power.

The question is: Can the spirit within the earth body converse with another spirit in earth body without regard to distance of separation; and, if so, under what conditions?

Answer.—The sublime science of spirit telegraphing is yet hidden in the laws of *action* and *reaction*, which pervade, and more or less obviously govern, all forms and gradations of matter. Every organ in the brain, and every ganglionic center in the visceral department, has its own peculiar sympathies. These sympathies are distinct and available. The superior organs generate exalted and expansive influences, which radiate over all other organs in the same body, and outwardly, also, to immense distances; and these influences are positively certain to touch and affect, in a similar manner, the corresponding organs and centers in other persons, whether they be absent or present.

The entire history of mankind demonstrates the foregoing principle. When a kingdom is involved in war, or when a people cherish combative and warlike feelings, the populations of both foreign and contiguous kingdoms are certain to realize a corresponding moral, and perhaps political disturbance. Thus, too, a frightful murder committed in one part of a great city, or small village, is almost certain to be immediately succeeded by several crimes of like magnitude and corresponding severity. Of rapes, suicides, accidents, robberies, treachery, falsehoods, mistakes, errors, &c., the same is undeviatingly and irresistibly true. The murder mania is thus explainable on philosophical principles. Not understanding this law of action, and reaction, an exchange says: "There must be some evil influence in the air, spreading murderous contagion, for murder has become epidemic. Besides the home tragedies that have recently appalled our city, the telegraph brings us intelligence of others perpetrated at the same time in various parts of the Union. Every day brings its record of fresh murders and attempts to murder, and it is a noticeable characteristic of the present bloody mania, that women and children are the most frequent victims." Now it is plain that the author of the foregoing remarks did not realize the law of social sympathy that underlies all social phenomena.

"Behold what a great fire a little matter kindleth!" A turbulent and wayward character in the habitation of angels, would excite like feelings, and provoke like conditions in the most amiable and lovely. Reason, on this law, appeals to reason, love to love, hate to hate, disease to disease, health to health, virtue to virtue, vice to vice, crime to crime; for the laws of social intercourse and brotherhood are universal, and it is folly of the sickliest kind to expect harmony in a part of the world, while all other parts are *telegraphing the messages of discord* into the finest deeps of the sympathetic soul.

Unlimited mental intercourse and social sympathy, therefore, are productive of either pain or pleasure. Effects will correspond to the generative conditions. We are desired to define the conditions. Nothing is easier, or more simple, or more certain of demonstration. What is true of two individuals, will apply equally to any two kingdoms or nations. The psycho-*telegraphic* law of *one* isolated soul, in the secrets of its own dual constitution, is the law of telegraphing between any two souls through any distance. What is the law, and what the conditions of its operations, in the individual? Briefly these :

Feet telegraph their sensations to the brain. There are hundreds of material obstacles and prominences between them, yet they sympathize and converse. Foot says: "I am lame and sore from over-walking." Brain receives the telegraphic message, and responds: "You shall be comforted." Foot replies: "Thank you—hope you'll keep your promise." In this familiar manner each organ converses with every other organ, and then they all, individually and collectively, report at headquarters—at the universally acknowledged seat of government—*the mind*, which is enthroned at the mountain top of all organizational existence.

What are the conditions? Manifestly these: that foot and brain be connected by some subtile cords of sympathetic contact. The same cords are necessary between all other parts and extremities. But how can these conditions exist between two congenial souls, "wide as the poles asunder," and in the external world? Thus: by a mutual understanding that, at a given hour of the day or night, when all the rest of the world is shut out of the charmed circle, each will *think a certain kind and number of thoughts with reference to the other*, with all that distinctness and earnestness which would naturally characterize a familiar face-to-face conversation. The amount of *time* to be consumed in thinking such thoughts, and the exact *method* of arranging them into sentences, or questions and answers, should be a matter of prior mutual understanding. Note down every thought that bolts in upon the mind while so telegraphing. In this way a melodious concert of sweet sympathies will be organized; after which, notwithstanding the immense distances, the twain may commune on the principle of the magnetic telegraph. We will cheerfully give more on these important points, if it be desired.

In one short sentence let us commit ourself to the long-cherished conviction that, in the not far future of this life, mankind will enjoy *telegraphic* intercourse independently of physical agents and machinery.

Nature's Progressive Energies.

Nature's heart is filled with forces and principles of perfection, and nothing can resist their ultimate manifestation. A strongly constituted man, for example, will recover from sickness *in spite* of blue pills and the lancet. So the whole body of mankind, being filled with every adequate energy, and with conquering principles, will make progress *in spite* of earthquakes, epidemics, bad religions, oppressive governments, and destructive wars.

It is true, however, that sections of Nature (below the spiritual Man) constitute a kind of War Department—a West Point Academy—where the quadruped brain (which yet remains in some men) acquires the art of living by means of violence and bloodshed. Life feasts upon death. Construction employs Major-General Destruction to superintend the progressive advancement of organic existence. That which lives in the world depends upon that which *dies* in the world. Destruction spreads the table for the support of construction. All departments of Nature, therefore, are regulated by the mutually operative wings of Progress, viz: Destruction and Construction, or Death and Life, or War and Peace. The bird eats the fly, the owl eats the bird, the hawk eats the owl, man kills the hawk, and so all the way up the steps of the organic growth; and yet we hold that Human Beings are not designed to be influenced and educated by *our inferiors*—by the fish, and birds, and animals, that live and breathe at the foot of the throne on which mankind sits—" a little lower than the angels."

Minerals, vegetables, and animals, climb up to the production and position of Mankind by means of force-won possessions. Beasts have war establishments in their brain, and teeth, and claws. Race eats race, as streams run into streams, to make the ultimate. Force and violence are natural, until the *spiritual* is reached; then the spiritual is the natural, and force and war are monstrous and unnecessary. Let each reader ask himself this question: "Is war congenial to my reason and affections?" If the spirit within shall whisper "yes," then blushingly and sadly we write the verdict, that your development *is not spiritual.*

Spiritual Briers and Thorns.

A certain correspondent addressed us as follows:

"SIR:—All I now require is a few direct, practical, inoffensive words. Can you explain to me the cause of my failure to interest persons, of either sex, in my feelings and most cherished views of religion. . . . . I am dreadfully cast down at times—hopeless, have thoughts of suicide, hate everybody, and everybody shuns me as though I was a nest of vipers, or a tree of poison and thorns. . . . Do give me a simple 'Whisper' of explanation, and I will remain forever obliged."

REMEDY.—Abolish from your mouth everything, whether fluid or solid, which is not necessary to meet the demands of honest hunger, and to quench the burnings of honest thirst. One hundred days from the commencement of this mode of life, you will begin to feel and to act like a new man. Your impatience will be diminished, your irascibility of temper and your overtaxed nerves will rapidly subside into peaceful conditions, and your excessive sensitiveness will depart from the surface of your external character.

Now begins the struggle between the Will of your awakened aspirations and the propensities of your inherited and acquired characteristics. Buckle on your whole armor, Brother, and prepare your will for a conflict with passion. Your wrestlings with the Satans of inherited discords and propensities will be sternest and most painful in your bed-chamber. Your disposition is hard to control. Sensitiveness is an effect of your diseased brain and nervous system. You are easy to imagine yourself "misused"—"slighted"—"insulted." O, how easy it is for you to think evil of your own best friends. All alone, in secret—where no human eye can see your frowns, and the bitter curl of your lips, and where no human ear can hear the low blasphemy of your wrathful mutterings—yes, in your bed-chamber! Thus, Brother, you convert yourself into a thorny tree of poison, bearing bitter fruit to those about you—a thistle bush, which children dread, and which even birds and beautiful animals will not visit. Everybody walks around you! People would go ten rods out of their way to shun your dark frown, and to avoid your complaining tone of voice.

And yet you think "folks are at fault" in all this, and intend merely to annoy and to offend you, because of your unpopular opinions. Not so. Love begets love, friendship enkindles friendship, good nature and civility awaken good nature and civility. Try it, Brother, and mark the fact! People will soon forget to hate you for opinion's sake. Everybody and their children will say: "We like M———; we love him dearly; he's *so* good and *so* cheerful; but we don't like his doctrines—that's no consequence among friends." We admonish you to contemplate subjects which will lift

> "Thy thoughts
> Far above the dust of worlds."

It is impossible to think or to act unworthily of your better nature without degrading the waters of life at their fountain head. Take heed! No more littleness: no more offending the image of purity in thy bosom; no more hostility to the practical teachings of Harmonial Philosophy; no more secret violations of the sweet processes of daily life; no more supposition that you can do evil, or think evil, and yet escape the consequences in all their magnitude; no more imagining that the sight of guardian spirits is dimmed by the walls of your chamber, or that you can be a brier-bush in social life, a poison tree in the garden of the family, a thistle in the hands of friendship, and yet be loved, and courted, and aided in business by your neighbors and acquaintances. Come, Brother! be simple and strong, sweet and healthy, intelligent and affectionate, spiritual, harmonious, hopeful, and free as the air of celestial mountains. When you think that you are worthless and unimportant, or when thoughts of evil sweep through your mind toward a fellow being, then read the following inspiration, by Bailey, the author of Festus:

> "Nothing is lost in Nature—so, no soul,
> Though buried in the center of all sin,
> Is lost to God: e'en there it works His will,
> And burns to purity. The weakest things
> Are to be made examples of His might—
> The most defective, of His love and grace,
> Whene'er He thinketh well. Oh! everything
> To me seems good, or tendencing to good;
> The whole is beautiful! and I can see
> Nought absolutely wrong in man or nature—
> As from His hands it comes who fashions all,
> With qualities in germ that shall unfold
> All holy as His word. The world is but
> A Revelation. His Spirit breathes upon us
> Before our births—as o'er the formless void
> He moved at first—and we are all inspired
> With His Spirit. All things are God, or of God."

Cause and Cure of Impatience.

"Intellectually speaking," says a reader, "I delight in the speculations and beautiful theories of Spiritualism—in them I am a sort of *connoisseur*, walking about, superficially it may be, like one in the corridors and halls of a royal gallery of immortal paintings. Yet every day, or whenever I attempt to fix my attention upon any one of the pictures, something seems to blind my eyes so that I cannot see. A feeling of impatience seizes upon my thoughts, so that I fear that I can never *realize* anything of the 'Harmony' prescribed in your writings as prerequisite to true spiritual enjoyment. . . . . . Why is this? Will you oblige me so much as to explain the cause of, and prescribe a remedy for, this impatience?"

REMEDY.—Corridors and halls of Art are frequently visited

by individuals who have in no way qualified themselves to study the triumphs of genius and inspiration. Many people look at a picture with such *haste* as to completely "blind their eyes" to its intrinsic merits. Many correspondents write to us for instruction in psychical laws, by which in a few weeks, or months at most, they may enter upon all the pleasures and spiritual benefits of the "Superior Condition." Such ambitious ones do not often remember that every Captain on shipboard served a term of years "before the mast," or among the "hands" by which the common work of life is accomplished. Experience makes perfect, and *time*, with obedience to the laws of personal improvement, brings the legitimate reward. Nothing is truer than the saying, that "The way of the transgressor is hard."

The writer of the above experience is a "transgressor." This will explain the failures of which he complains. In approaching the "beautiful theories" of the Harmonial Philosophy, and while attempting to fix his "attention," his thoughts fly off like sparks from the blacksmith's anvil. He finds himself a sort of "connoisseur," and nothing more, and then behold the consequences. He is deprived of the interior enjoyment which he intellectually sees hanging in luxurious profusion through the halls and corridors of the Spiritual gallery of Inspiration and Reason. All because he is a "transgressor." Of what? Of the psychical laws of QUIET. His motive and mental temperaments push him *beyond the tranquil point*, making meditative study out of the question, and of course "a rolling stone gathers no moss." He is nervous, hasty, precipitous, in desperate haste, in a flutter, and feels hurried; and, neglecting to analyze the sources of his impetuous impulses, he is more than half inclined to be *superstitious*, like a goodly number of chaotic minds, and charge his lack of concentration at a particular point to the intervention of "spirits."

Friendly reader, see to it that you do not superstitiously trace your mental impatience to a false source. Believe us, the cause of your defeat is nestling in the *unbalanced* condition of your temperaments. When you would do good, evil (or chaos) is present with you. Cultivate the Will-power, and hold your thoughts to their purposes. Give yourself more *time* to enjoy a truth. Drink deeply, but with great deliberation— always taking *a long breath* of meditation between each draught, while at the flowing spring of spiritual realities. Divine ideas require very respectful attention. Speed in spiritual things is impossible. It takes the best parts of a year to raise a moderate harvest of good thoughts. In the cold and stormy "winter of discontent"—which every soul is certain to experience—such harvested thoughts will feed and nourish the INTERIOR.

Od-Force and the Odoscope.

The etymology of these terms has been given by the learned Reichenbach, in the following translation by Hittell: Od is a force analogous and nearly related to the other forces already known to science. It includes a group of natural events, (*Vorgaenge*,) improbable but perceptible to the senses, for which we have no measure or agent save the human senses, and even these only under peculiar circumstances of the sensitive impressibility.

The reason why it has hitherto escaped scientific investigation, and has even been directly and stubbornly repelled and excluded by science, lies in the want of a universal odoscope or odometer, which might be placed within the reach of every one, and whereby its existence might easily be demonstrated before the eyes of all the world.

And again, the reason why no odoscope has yet been discoverable, springs from the nature of od itself, because of its power to pervade all things and all space, without accumulating in any place sufficiently to become perceptible to mankind generally. There are insulators for heat, electricity, and light, but I have been unable to find any for od. This want of confinability, I have thought proper to use as a hint for a name which might be suitable for the varied combinations of scientific nomenclature.

Va, in Sanscrit, means to blow. In Latin *Vado*, in old Norse *Vada* means I go quickly, I hasten away, I flow. From that *Wodan*, in old German, means the all-pervading; it changes in various dialects to *Wuodan, Odin*, signifying the all-pervading power which is finally personified in a German deity. 'Od' is consequently the name for a force, which, with irresistible power, rushes through and pervades universal nature.

Physics and Metaphysics.

PROPOSITION.—Metaphysicians sometimes present man to us as a mere machine, and at times as the soul of all machinery, or even the great Mechanic. Now, "who shall decide when doctors disagree?" But is it really so difficult to show what part of the divine drama he plays? Cannot his whereabouts be defined to the common understanding to know whether he is an individual of duty and responsibility or not?

There is certainly a vast deal of darkness, uncertainty, and error, in relation to this, among those who have joined the reform discipleship. Some consider themselves *machines*, and will rest in quiet till they are operated upon; while others consider themselves souls, or mechanics, and are ever positive and active, turning the world upside down in *reform* or *deform*, they little know or little care which. If men are machines they are not responsible; if they are mechanics, they are.

The above opens a great and important question. If mind

can act in the absence of predominant motives, he is a mechanic; if not, he is a machine. And it either *can* or *cannot;* it is an absolute alternative, and no law can fail to have a direct bearing upon the yea or nay. I therefore submit this question: *Can mind act in the absence of predominant motives?*

EXPLANATION.—The most skillful navigator on the mystic sea of metaphysics will sometimes misguide his ship. He occasionally will, with the best phrenological chart, and with the best mental compass, steer into quicksands, and founder in sight of port. We have witnessed sad shipwrecks upon rocks with which far less intelligent pilots were perfectly familiar. Time never was, and the day will not soon arrive, when every man will escape the perils of progress. The storms of discord are natural, and the disasters consequent upon them are natural also, and we compassionate and pity the mind that would fight against Nature.

Now we say, and with due reflection, that certain men, with certain temperaments, are natural fatalists, and such are, by necessity, quite *mechanical* in nearly all their thoughts and conduct. They are fixed fast in the wheels of incorrigible destiny. They move with the motions of the world. The laws of cause and effect—of Necessity—are, to such, clear as the principles of mathematics. They are spiritual Necessitarians, natural machines, fatalists, presbyterians; and there is no honest philosophy, no true logic, no natural religion, that can drive them from their strong fortifications.

Why not? Because they have not only the inexorable laws of logic to sustain them, and to explain their positions but *experience* also; so that, as a correspondent truly remarks, "there are things in man's most intimate economy which perfectly stump the imagination."

Especially so, because there are other men, with other temperaments, who feel an integrality of indestructible powers;

who do not act from sensuous impulses; who receive and yet control the influences by which others are overcome and compelled to act; who give fashion and shape to the circumstances about them, just as the potter molds the moistened clay. They work from *within*, and their action is in consequence of self-conscious and self-authorized impulses. They believe in "free will," and in self-originated motives for conduct; and they laugh at those who profess to act by *necessity*, in obedience to fixed laws.

Who will answer our correspondent's question: "Can mind act in the absence of predominant motive?" Or, is not action by man a result of some power superior to his will? Is not his will a *subject*, not a master, in the circle of existence?

Our reply (how unsatisfactory!) is YES, and NO. Different temperaments, with different experiences concerning the same matter, will entertain different convictions. Some are masters, while others are slaves, within the same set of circumstances. Human experience and observation, therefore, are ample on both sides of this question; and the consequence is, that doctors differ, and erudite metaphysicians disagree; for Nature and Truth are on both sides of this question also.

And we experience no conflict when so contemplating the answer. We perceive that man (*i. e.*, mankind) is both a Latitudinarian and a Necessitarian. Not every man, indeed, but only he who is rounded out. The true man is self-poised, self-intentional, and grandly responsible for his conduct; and yet, here comes the paradox, he is perpetually obedient to the fixed laws of the Infinite.

The human individual's responsibility is commensurate with, or in proportion to, the mind's power to conceive of justice and freedom. He who seeth the way to do better, and yet goeth not therein, moved thereto by his love of justice, is responsible to the Divine presence which worketh within both day and night. Let the gospel be full-spread everywhere, that Man, in the wondrous duality of his being, is forever a master over conditions, but a subject of the laws by which those conditions are generated. That is to say, Man is subject to the law of digestion, but he is, or should be, master, with respect to the *conditions* and kinds of food.

Marriages of the Temperaments.

When quite a boy (says a correspondent,) my father was born into the Spirit World. Since then I have lived in *sixteen* different homes. Have seen a very few happy and orderly families. The many are full of "angles," discords, and contentions. I see two great causes of all this: first, we come into the world deformed, either physically or mentally (caused by an ignorance of the natural Laws of reproduction;) second, the nutritive-sensuous temperament tries to marry and live with the motive-mental temperament (caused by an ignorance of the natural Laws of marriage.) Now, I ask, which of these two causes is the *most* prolific in depriving the married of true happiness? I have seen the lower temperaments live happily together, while I know of no instance where the lower and the higher temperaments were happy in marriage. The first feed upon "pork;" the latter upon "thought."

EXPLANATORY REPLY.—We deeply feel the responsibility which attaches to whomsoever assumes to analyze and treat upon this forbidden subject—"marriage." There is certainly no relation of profounder import, of wider range, more delicate, or influential on human affections, prejudices, and destiny, than that to which our observing correspondent asks our attention. Having lived through all the life of sixteen families, felt their discords, relished their contentments, and observed from the Harmonial standpoint the many and diverse *causes* of their unhappiness, he comes upon the stand as a witness, in the full use of all his senses, declaring that one of the "two causes" prevails in every case where suffering and discord gain the ascendency over peace and harmony. And his question is, "Which of the two is the most prolific source of trouble among the married?"

Our reply in this connection must be brief, though the theme demands lengthy elucidation. And in our answer we include all professional persons of every country, laborers, mechanics, rich and poor, philosophical and uneducated, because the Central principle is not partial and unjust, but universal and everywhere applicable—viz.: *that blood-love* (which is a reproductive passion,) *and spirit-love* (which is a divine attraction,) *can never assimilate and dwell harmoniously in the relation of marriage* In strictest truth it may be affirmed that no human or so-called supernatural authority can ever join such temperaments in true nuptial relation. They live a tedious life, compounded of misery, detestation, bitter mockery, and what is worse, doing a cruel injustice to the consequent progeny. Priests may perform at the marriage altar, and say, "What God hath joined together, let no man put asunder," but Nature and Nature's God say, "These persons are not one, but twain, being unmated in essence, wherefore no power in heaven or on earth can pronounce and crown them as 'husband and wife.'"

The true and effectual remedy, as we have many times affirmed, is a universal knowledge of and obedience to the natural law of temperamental adaptation. Legalized unbinding, or divorcement, according to the decrees of statute law, though in many cases absolutely necessary and just, is, after all, but treating effects, palliating the pain of consequences, while the sources and producing causes in society remain in full supply and energy. We say, then, incorporate a knowledge of such temperaments as may be joined in wedlock, with every young person's education; thus multiply the number of right marriages, of happy homes, and diminish the sum of human misery. (See fourth volume of the "Great Harmonia.")

Progression of Primaries.

The philosophy of the progressive development of primaries, through proximates to ultimates, is peculiar to the Harmonial Revelations of Nature. The hypothesis cannot be traced to any other form of truth. We hold that all matter goes *an infinite number of times* through every link of the endless chain of progression. It flows back and forth, down and up, in and out, without cessation, from the commencement of one eternity to the birth of another, before it reaches the highest form and sphere of organic life—*i. e.*, the physical body and spiritual constitution of MAN. We are delighted with the latest chemical discoveries. They bring out and establish among scholars, and, therefore, in the schools of our land, the very principles for the perception and advocacy of which the sectarian portion of the world labels us with unbrotherly names.

Law of Contagions.

Sympathy, or contagion, is the general law of human nature. Every act of a human being is referable to some particular organs of mind. These faculties emit an influence which acts correspondingly upon the like faculties of other persons. Hence, vice, as well as virtue, is contagious. Those who live in the midst of fraud, poverty, vice, and profligacy, necessarily absorb the degrading nervauric magnetism thus generated, and lose some of that higher influence which stimulates and builds up the superior faculties.

Exhausted Primates in Man.

A Philadelphia correspondent sends us the following curious diagnosis, which he received from a medical practitioner, who resides not far from the "hub of the universe." The patient, writing for information on some obscure point of the diagnosis, says : " Some six weeks ago a person, in Boston, had a flaming advertisement representing himself as being able to cure very difficult cases of disease pertaining to males and females. I wrote to him, and stated my case in full. I will give you some extracts from his letter, viz.: ' Cold water never alone could cure you, for you, as well as all other human beings, are compounded of some seventy-five primates, one or more of which has become exhausted, and, therefore, must be restored before you can regain perfect health. In your case, the absent primates appear to be, 1st, Lime ; 2d, Manganese ; 3d, Phosphorus ; 4th, Iron ; restore these to the system (after treating the chronic inflammation of the parts involved,) and you must get well at once.'

"He then describes the treatment which is necessary, and adds, 'This will cure you perfectly, just as surely as that one and one make two.' I did not try this treatment, as his charge for furnishing the medicines was fifty dollars."

ANSWER.—Our correspondent wants us to solve many questions concerning this doctrine of the "primates"—whether they consist of metals, salts, &c.—and whether man must take "cordials," &c., in order to keep his "primates" in sufficient quantity and equilibrium. We answer, that the learned words at the head of this paragraph, contain considerable intelligence when applied to soils or plants ; but they mean absolutely *nothing* when applied to the constituents of the human constitution. The diagnosis above given is without foundation in Nature. It is simply unsound and ridiculous in the eyes of Science ; and the theory is *dangerous* to the millions who know little of the chemistry of bodily existence.

All substances in the world are composed of sixty-four (not seventy-five,) simples, called "primaries," because we first find them in rocks. These rocks, by means of pulverization during the labor and lapse of ages, result in soils. From these soils vegetables are unfolded, which lift up and still more perfect and refine the "primaries," until they become sufficiently attenuated and potentialized to unfold and sustain the organization of animals. Man's constitution is a reservoir for all the *ultimates* (or proximates,) of rocks, soils, vegetables, and animals. He does not exist nor subsist on the primaries. The basic elements are first taken up by the lower orders of plants; they progress through all the ascending grades of vegetable bodies, till they form part of the air, and water, and food of animals; and, still passing gradually upward, they (the primaries,) *ultimate* in the human organism. Man, therefore, is composed of ultimates; not of crude "primates."

In view of this truth, how shallow—not to say mischievous— is the above diagnosis! The patient must take medicine compounded of gross and indigestible primaries! Rocks, metals, earthy matters, must be dissolved or mixed with liquor and syrup, and then spooned out to the unfortunate victim! Chemistry of itself will expose the fallacy. The higher orders of plants will not appropriate the *crude* properties of the sixty-four simples. It is necessary to feed plants with the manures of other plants, or with fertilizers that have been refined by passing hundreds and hundreds of times through the life-processes of lower orders of animals. For this reason, our intelligent farmers and horticulturists find valuable manures in fish, in pulverized bones of animals, and in lime composed of the infusorial remains of departed ages.

The truth, then, is precisely and unmistakably this : Man is compounded of *ultimates* (which were once contained in the primaries and proximates,) and it is, therefore, impossible to strengthen him, or to cure his maladies, by dosing him with crudities and mineral masses; for these all are foreign, incompatible, and consequently disadvantageous, if not poisonous, to his physical and spiritual constitutions. If all readers will, from this hour, obey the simple laws of life and health—as to diet, sleeping, drinking, exercise, breathing, willing, and magnetism—they will find all deficiencies of body, and even the evils of transgressed laws, gradually supplied and transformed to good. "Throw physic to the dogs," scorn the chicanery of advertising pretenders, and at once set up for yourself. This prescription will cure you as certainly as that " two and two make four," and our fee is *not* fifty dollars.

Sources of Vegetation.

The vegetable kingdom is composed of materials derived from the mineral kingdom, which is hidden in the earth's deep bosom. Whence the materials composing the mineral world? From the four primal elements of all matter—viz: Fire, Heat, Light, and Electricity.

In all searchings, we terminate in these four original principles. By "Fire" is not meant the condition of matter in flame or combustion, but the *finest state* of material motion, out of which issue Heat, Light, and Electricity. The finest motion of matter results in the illumination of matter from its own inward sources. Electricity of immensity is the conveyance of all vital action in the universe. Fire, Heat, and Light, are passengers in the omnibus of Electricity. They get in this chariot of infinitude, and the integral motive-principles propel it through every avenue of boundless Nature.

The omnibus stops for a moment at the different stations along the interminable highways of infinity; and, at such a moment, passengers enough leave it to start the organization of a *world* like this globe on which we live. Fire, Heat, and Light —the three grand primal principles of matter—retire from the Electrical chariot in minute particles, and forthwith unite, chemically, for the origination of a planet. [The first stage of a planet is invariably that of a comet.] Then onward rolls the triumphant chariot of Electricity! Passengers are permitted to terminate their ride at just that point in space where they experience more attraction to sojourn than to proceed. But it is very strange, yet true always, that only a sufficient number of passengers (or particles,) alight from the shining vehicle, to commence the new planet in that particular locality.

You perceive that Fire, Heat, and Light—the primary principles of all matter—contain the requisite properties for the composition and ultimation of all mineral, vegetable, animal, and human bodies. Of the life-forces and spiritual sources of these different bodies, we do not now say anything; believing that, with the foregoing sketch of the origin of planets, and of their elemental primaries, you can settle many other problems to your perfect satisfaction. Think, contemplate, and you will see

> "How Father God,
> From the mute shell-fish gasping on the shore,
> From men to angels, to celestial minds,
> Forever leads the generations on
> To higher scenes of being; while supplied
> From day to day with His enlivening breath,
> Interior orders in succession rise,
> To fill the void below"

How to Balance the System.

PROPOSITION.—" You say, 'Harmony is possible only when both brains, back and front, are equally exercised.' Physicians have told me that my continued poor health was the result of an undue activity in the frontal lobe of the brain, which robbed the rest of the system of its vitality, and that I must *stop thinking*, and cultivate other faculties or departments of the brain. Very likely this is good philosophy, but the question is, *how to do it*? If my health depends upon the answer to this question, it certainly is a *very important one.*"

ANSWER.—Your physicians are scientifically right. An

undue activity in the voluntary or frontal lobe of the brain, is certain, imperceptibly, to refine, spiritualize, and attenuate the sanguineous and muscular systems; but physiologically viewed, the opposite effect is certain to be a corresponding amount of debility in the vital organization, resulting in a variety of functional weaknesses and prostrations, wholly unfitting both mind and body for cheerfulness, naturalness, and even ordinary beauty of expression. The philosophy of all this we cannot stop to present, but will at once define a few simple rules in favor of restoring an equilibrium.

First. Observe, with cheerful gratitude, the *taste*, the flavor, of everything you eat. Think of its *odor*, also, and endeavor to appreciate, *by feeling*, the goodness, and the virtue or sweetness thereof. Acquire the habit as *soon* and as *perfectly* as possible. (Of course, you will not eat or drink any substance, or fluid, unless its *flavor* and *odor* are wholly congenial.) The habit of thinking upon *foreign* subjects while partaking of food, is very unphysiological, and even wicked. We have oftentimes, while dining with a friend, been interrogated upon some abstract subject; but our answer has uniformly been, " *one thing at a time.*" In truth, no mind can be illuminated while in the proper act of feeding the physical organization. *Be entertained with your food.* Social converse, with natural feeling, and the mirthful chattings which naturally come of it, make business enough for the brain while eating. The ghostly, stately, ministerial, iceberg dignity of some believers in old theology, during the happy process of food-taking, is hurtful to both body and mind, and will be rebuked by every physician whose brain is sealed with the crown of common sense. *Be genial, be good, be grateful!* these are the only infallible anti-dyspeptic medicines; the only perfect remedy for torpid liver and constipation.

Second. Never read upon any deep philosophical subjects

after twelve o'clock, A. M.; nor upon any occult, abstruse, or spiritual question, after four o'clock, P. M.; nor attempt to meditate in the regions of sublimity, ideality, conscientiousness, or veneration, after the evening's repast. None of these rules are positively applicable, except in cases where the cerebrum, the front brain, has gained the supremacy over the vitalic forces, and has, in consequence, become *involuntary* in the performance of its thought-generating functions. This almost every intelligent physician will tell you.

Third. In regard to food, clothing, and exercise: You require very little fluid of any kind. Solids from either kingdom, the vegetable or animal, in small quantities. Hunger is indispensable to gustatory enjoyment. If you do not get truly, stoutly, and universally hungry—so that every mouthful is a grateful luxury at the time—then wait, go without both solids and fluids, until Nature's infallible voice, "slay and eat," is heard emanating from every joint and organ. Then obey; but only as you would feed an infant. The whole digestive system will "cry for more," doubtless; but give to your mouth *feelingly and temperately.* The divine God-code of laws is written all over the membranes of your mouth and digestive system. Believe—obey—be grateful—and thus convert every meal into an act of worship—just as the bee gathers honey, as the bird sings, as water makes music, while performing innumerable missions in Nature.

Do not be *too conscious* of existence by means of thinking. Some persons think, and think, and think, until they think themselves into a sickly nervous fluid. Such become intensely aware of self—of individual feelings—of besetting wants and needs—of egotistic necessities, which many times become imperative and uncontrollable. All this is disease. Feel nobly, and think only as you feel, except in the morning hours, when it is right and healthful to *think*, and *read*, and *study*, and *contemplate*, whether your feelings coincide or not. Let every one, young or old, sick or well, rich or poor, try to acquire these intellectual, these social, these harmonial habits of life. They will bring healing in their wings.

Of Clothing: Always wear next to your flesh what is most congenial to the cutaneous sensibilities. Protect weak and sensitive places with more garments than you wear upon healthy parts of your body, and do this in opposition to whatever "fashion" may prevail in the surrounding world of dress. At another time we propose to take up this dress-question, and will, therefore, abstain from further remarks in this connection.

Exercise should always be directed to reach and *strengthen* the weakest and most undeveloped parts of the body. A private gymnasium is less expensive than a silk dress. A whole family may be benefited by an outlay per annum about equal to the cost of tobacco for the masculines during that period. Long walks are, many times, worse than no exercise whatever. Horseback riding in the afternoon, or driving about in an open carriage, is much better than long walks—especially the former—yet, as the human constitution is made for every imaginable variety of motion, no *one* exercise can ever comprehend and stand substitute for all other kinds. And remember, that what we term "Rest," is not inaction, but a *change* of occupation; a different form of motion, another employment of the same muscles and identical faculties.

Finally, dear friend, we admonish you to begin—although this may be a late day in your earthly life—*to be harmonial and natural.* Untrammeled by the demon-influences of old theology; free from the yoke of bondage to the dismal superstitions of popular creeds; open to the warmth and light of infinite love and wisdom, as displayed all around you, as incarnated even in your own personal organization—we counsel you, in the name of all these blessings, to systematize your daily habits of *feeling* and of *thinking*, so that a beautiful healthfulness may be conceived and brought forth, as the beginning of your everlasting salvation from discord and unrest.

An Orange before Breakfast.

QUESTION.—" DEAR SIR : As I am always for knowing the *why* and *wherefore* of things, you will confer a great favor on me if you will state the reason why you prescribe that a person *should eat an orange in the morning, when walking out early, and should, after returning, go to bed again.*"

EXPLANATION.—The medicinal properties of *citric acid*, or the acid of lemon, are positive and useful in several conditions of the system. But we prescribed the milder form of acid to be derived from the *citrus aurantium*, which is the Latin for Orange. If the acidulated juice of orange be consigned to the stomach while empty and weak, as it is in the morning, the effect, on a few trials, will be exhibited in the disappearance of bile (alkaline matter,) from the duodenum, a revivification of the digestive powers, and a cessation of nervous headaches. But neither of these desirable effects will follow the eating of an orange before breakfast, unless the other directions are punctually adhered to, especially the walk and *the sleep* previous to taking the customary substantial food of the morning.

The reason why sleep is particularly necessary after a morning walk, and before breakfast, is, the *restorative influence* which a repose of blood, bone, and brain, at that hour, is certain to exert upon the entire organization during the rest of the day. On the other hand, active employment of the brain, or muscular labor, or long walking in the early morning—without a rest and sleep after such exercise, and before eating—is quite likely to result in an increase of digestive prostration, considerable headache, greater nervous irritability, and a more general lassitude. The orange juice is gratefully refrigerant and cooling to the heated membranes of the dyspeptic stomach, distilling a fine positive power upon the pneumogastric centers, and preparing the blood-making system for the healthful discharge of its extra-important offices. (We say all these things, remember, as applicable, chiefly, to the Debilitated.) The sleep remedy is of the greatest consequence in debilitated conditions of the blood and nervous system. Consider the ability to sleep after a morning walk, and before eating nutritious food, as an accomplishment of great individual merit. It will bring you an abundant reward—peace, refreshment, satisfaction, health.

Early Rising Triumphant.

Miss Martineau is a great advocate of Early Rising. She says: "I speak from experience here. For thirty years my business has lain in my study. The practice of Early Rising was, I am confident, the grand preservative of health, through many years of hard work—the hours gained being given, not to book or pen, but to activity. I rose at six, summer and winter, and (after cold bathing,) went out for a walk, in all weathers. In the coldest season, on the rainiest morning, I never returned without being glad that I went. I need not detail the pleasures of the summer mornings. In winter, there was either a fragment of gibbous moon hanging over the mountain, or some star quivering in the river, or icicles beginning to shine in the dawn, or, at worst, some break in the clouds, some moss on the wall, some gleam on the water, which I carried home in the shape of refreshment. I breafasted at half-past seven, and had settled household business, and was at my work, by half-past eight, fortified for seven hours' continuous desk-work, without injury or fatigue."

Objections to Early Rising.

PROVIDENCE, R. I., June, 1860.

DEAR SIR: I have often read with interest your medical productions, and, as far as actual effort has gone, think I have derived material benefit from your disclosures and suggestive declarations. But will you not dissect the *following*, against your theory of Early Rising, which may be found in *Hall's Journal of Health*, volume 7, page 116. He says:

"Breakfast should be eaten in the morning, before leaving the house for exercise, or labor of any description; those who do it will be able to perform more work, and with greater comfort and alacrity, than those who work an hour or two before breakfast.

"Besides this, the average duration of the life of those who take breakfast before exercise or work, will be a number of years greater than of those who do otherwise.

"Most persons begin to feel weak after having been engaged five or six hours in their ordinary avocations; a good meal reinvigorates, but from the last meal of the day until next morning, there is an interval of some twelve hours; hence, the body, in a sense, is weak, the stomach is weak, and in proportion, cannot resist deleterious agencies, whether of the fierce cold of midwinter, or of the poisonous miasm, which rests upon the surface of the earth wherever the sun shines on a blade of vegetation or a heap of offal.

"This miasm is more solid, more concentrated, and hence, more malignant, about sunrise and sunset, than at any other hour of the twenty-four, because the cold of the night condenses it, and it is, on the first few inches above the soil, in its most solid form; (but as the sun rises, it warms and expands, and ascends to a point high enough to be breathed;) and being taken into the lungs with the air, and swallowed with the saliva into the stomach, all weak and empty as it is, it is greedily drank in, thrown immediately into the circulation of the blood, and carried directly to every part of the body, depositing its poisonous influences at the very fountain-head of life."

Now, if this is really so, your many readers ought to know it, and act accordingly. Yours, for truth, E. T. D.

ANSWER.—We can perceive no "poisonous miasm" a few inches above the soil, except in regions where large bodies of vegetation and stagnant waters are undergoing rapid decomposition. The ordinary reduction of temperature, by which dew is condensed from atmospheric humidity, or the corresponding elevation of the temperature, by which the same dew is returned to the upper air, are *changes* more to be provided for and shielded from, than any of the miasm consequent upon them The true external protection is *dress ;* while *food* is naturally the safeguard of the inner man ; in addition to which may be mentioned *exercise*, and the active breathing inseparable therefrom.

All the moving millions of lower animated Nature—the birds and worms, the quadruped world, and almost every form of vegetable life—follow the solar laws with instinctive precision. They seek repose and recuperative slumber when the sun sets, and with auroral light they awaken to the duties of the day. The period of magnetic attraction commences in our latitude positively at the moment of full sunrise, and although it is chemically true that unvitalic ethers rest upon the soil, both in the evening and in the morning, yet is man physically capable of separating the good from the imperfect, and that, too, without making the process a matter of design.

The civilized method is, judicious clothing and sufficient feeding of the body, which is the chemical factory for the elaboration of the form in which we shall hereafter appear. Man should not trouble himself about fogs and miasms in this latitude. With regard to Early Rising, or exposure to the evening air, the whole physiological law is fulfilled by attention to one point—TEMPERATURE. Let this vulnerable point be well guarded, and all the rest will redound to his advantage. It is demonstrably certain that all human diseases originate in a reduction, or in too sudden and too frequent alteration, of the bodily temperature. Keep this temperature as nearly as possible in a state of equilibrium, or guard against too rapid transitions, and you are almost absolutely impervious to the encroachments and bombardments of disease.

In keeping with this philosophy of health and disease, we prescribe for the "Debilitated" an early morning ramble, the eating of an orange, at least thirty minutes of quiet sleep, and a light hand-bath—all previous to the first meal. In many cases the luxury of sleep is unspeakably better for health than eating.

The French proverb: "He who sleeps, eats,"—is the verdict of experience. For digestive derangements and nervous diseases, where the brain is involved, we know of no food so good as an abundance of dreamless sleep.

We agree with Dr. Hall, that much exercise or protracted labor before breakfast is unphysiological. It will eventually prostrate the vitalic powers of the finest constitution, and shorten the natural period of earth-life. We do not, therefore, recommend much labor before the first meal. But we do enthusiastically urge the practice of "early to bed and early to rise," with a substantial meal to start the duties of the day upon, as the most rational and harmonizing life for intelligent human beings to live. And we believe that all who receive the principles of Harmonial Philosophy, will be willing to obey the voice of universal Mother Nature, when she calls her human offspring through the myriad voices of the lovely landscape, and through the golden light that comes streaming with such rich luxuriance from the regions of the far-reaching Aurora.

The doctrine of late to bed and *late to rise*, will meet the wishes and suit the habits of the night-loving population of gas-lit cities. It is very delightful to believe, after excessive midnight debaucheries, or with the fatigue consequent upon voluptuous indulgences, or after late novel-reading and intemperate habits, that the earth is covered with a "poisonous miasm" about the period of sunrise. It is so pleasant to sleep on, with the assurance of medical men that it is *unhealthy* to rise early with the singing birds! Nay, not so. Be up and doing while "the day lasts," for the earth is then filled with the innumerable splendors of the infinite Father.

Physical Strength and Energy.

Physical strength is the primal inheritance. Men could conceive of the principle of power when they could not comprehend the spiritual quality of love. It is natural, therefore, in a certain stage of human development, to believe in giants, heroes, and in gods of boundless powers. Great men, because good and wise, are remembered and venerated only by the cultivated *few* who appreciate spirit and cherish ideas. But mighty men, because physically ponderous and strong, are the historical companions and crowned gods of the millions. The *glory* and *power* of God were revealed long prior to the finer and higher attributes of *love* and *wisdom*

It is, consequently, natural for children to enjoy stories of the feats of huge giants—marvellous tales of some great strong man who could carry the earth upon his shoulders, or eat it, as some legends say the first men did. In keeping with this infantile law of growth, the first entertainments and amusements were physical; the first memorable exploits also; and the gods were promoted and adored in proportion to the demonstration which they made of skill, cupidity, and muscular supremacy.

Thus Hermes, or Mercury, represented as a pleasing young man, with bright eyes and a cheerful countenance, became the universal favorite. He was the champion of the physically strong during a considerable era in human history. Believing parents would relate his mighty deeds and crafty exploits to their children. He roguishly purloined the sword of the great god, Mars; took many ponderous tools from the blacksmith shop of the grizzly Vulcan; stole the girdle of Venus, and the scepter from the mightiest god, Jupiter; killed the hundred-eyed Argus; fastened the strong Ixion to a wheel in the infernal world; sold the powerful Hercules to the Queen of Lydia; delivered the mighty Mars from the superior power of Alcides; and for these exhibitions of physical energy and skill, Mercury was appreciated, promoted, and worshiped.

It is natural, we repeat, to believe in gigantic energies, and to admire extraordinary feats of bodily strength. We respect great physical power in man, as we venerate noble mountains in Nature. It is respect and fear; not love. Pigmies are oftentimes both pretty and pitiable; but giants are at once fearful and a refuge of strength; and to this conclusion the intellectual world is rapidly hastening.

Swimming schools and gymnasiums are developing on every hand; ball playing in summer and skating in winter, for the young or adult of either sex; and innocent games or entertainments, not injurious to good morals, but promotive of intellectual power, are multiplying in exact ratio to the invention of labor-saving machinery.

We were encouraged and delighted, as well as astonished, when authentic news came that Dr. Winship, of Boston, had succeeded in lifting the enormous weight of *one thousand, one hundred and thirty six pounds*, and that he did not doubt his muscular ability, one of these days, *to raise a ton from the earth!* The extraordinary feat inspired us with unspeakable gratitude. We realized more than ever the omnipotency of truth and mind over the material muscle. *It was all attributable to the wellspring of spirit and power within the human soul* --for every philosopher knows that *matter*, of itself, is powerless and dead, depending every instant of time upon the mental

15

forces for motion, life, energy, and government. So, too, when we heard that the New York gymnast, Ottignon, was enlarging his establishment, and improving his system of physical training, and that similar healthful institutions and enterprises were being inaugurated in other cities, for the physical benefit of pupils of either sex, we rejoiced exceedingly—for we believe in bodily perfectibility as a foundation of true moral development.

Power-Generating Habits.

We will give you a portion of the interesting testimony of Dr. Winship, of Boston, whose experience in power-generating habits is good authority:

"I was nearly seventeen years of age," he says, "before I seriously undertook to improve my physical condition. I was then but five feet in hight and a hundred pounds in weight. I was rather strong for my size, but not strong for my years, and my health was not vigorous. I am now twenty-six years of age, five feet seven inches in hight, and one hundred and forty-eight pounds in weight. My strength is more than twice that of an ordinary man, and my health is as excellent as my strength.

"What has produced this astonishing change in my physical condition during the last nine years? I will attempt to sum up a few of the proximate causes that may have led to this result.

"I have breathed an abundance of pure, fresh air almost constantly.

"I have exposed myself sufficiently to the sun.

"I have eaten an abundance of wholesome food.

"I have drank less than a quart of spirituous liquors, and less than a gallon of fermented.

"I have used less than an ounce of tobacco.

"I have taken, nearly every day, about half an hour's gymnastic exercise in the open air.

"I have conformed to the customs of society only so far as they were not at variance with health."

* * * * * * *

Among other important conclusions, we find the following:

"That increase of the muscular power was attended with increase of the digestive.

"That one means of increasing the digestive power was to increase the muscular.

"That many articles of food had formerly proved injurious to me, not because they were unwholesome, but because I was unable to digest them.

"That a person may become possessed of great physical strength without having inherited it.

"That, by increasing the strength, a predisposition to certain diseases may be removed, and diseases already present, removed or mitigated.

"That increase of strength cannot long continue on a diet exclusively vegetable.

"That increasing the strength made excretion take place less from the skin, but more from the lungs and the other emunctories.

"That what benefits a part of the body, benefits, more or less, the whole.

"That long before I succeeded in lifting 1,100 pounds with the hands, or in shouldering a barrel of flour from the floor, I had ceased to be troubled with sick headache, nervousness, and indigestion.

"That a delicate boy of seventeen need not despair of becoming, in time, a remarkably strong and healthy man."

Preparation of Healthy Food.

Dr. Trall, in his excellent pamphlet, has expressed numerous valuable thoughts on Dietetic Rules. "After all," he remarks, "the most difficult part of the Hygienic system is the management of the dietary. Few persons know anything about hydropathic cooking; and so perverted are the appetences of the masses, that to talk to them of physiological victuals is very much like talking to a brandy-toper of the beauties of 'clear cold water,' or to a tobacco-smoker of the virtues of a pure atmosphere. Bread, which is, or should be, the staff of life, has, by the perversions of flouring-mills and the bakers, become a prolific source of disease and death. Much as the Health Reformers declaim against the abominations of pork, ham, sausages, and lard, as articles of human food, I am of opinion that fine flour, in its various forms of bread, short-cake, butter-biscuits, dough-nuts, puddings, and pastry, is quite as productive of disease, as are the grosser elements of the scavenger swine.

"Nearly all the bread used in civilized society is made of fine, or superfine flour, which is always obstructing and constipating, and which is deficient in some of the most important elements of the grain; and it is still further vitiated by fermentation, or by acids and alkalies, which are employed to render the bread light.

"Pure and wholesome bread can have but three ingredients—meal, water, and atmospheric air. The water is only useful in converting the meal into dough; and the atmospheric air serves to expand its particles, so as to make light and tender bread. If properly managed, bread can be made as light as ordinary loaf-bread with no other rising than atmospheric air. To effect this, three essentials must be regarded. 1. The dough must be mixed to a proper consistence—neither too stiff nor too soft. 2. The dough must be cut or rolled into cakes, or pieces, so as to expose the greatest possible surface to the heat of the oven. 3. The oven must have a brisk, or quick fire."

In keeping with the foregoing *three* essentials, the Doctor gives a number of healthful directions, several of which we quote. We commend them to our readers generally, whether sick or well, students or farmers.

BOILED GRAINS.—Wheat, rice, hulled corn, and samp, boiled until the kernels are entirely soft, but not broken nor dissolved, rank next to bread in wholesomeness. They may be eaten with syrup, sauce, sugar, milk or cream, or fruit.

WHEAT-MEAL CRISPS.—Mix the meal with water, cold, warm, or hot, into a stiff dough; roll it out as thin as possible, and cut into small narrow pieces, or strips, and bake in a quick oven. These are excellent for sour stomachs and irritable bowels.

RYE-BREAD ROLLS.—Rye-meal (unbolted rye-flour,) may be made into bread-rolls, or batter-cakes, in the same manner as for wheat-meal bread rolls. They are very light and delicious.

LOAF-BREAD.—Very good loaf-bread may be made of six parts of wheat-meal, two parts of corn-meal, and one part of mealy potatoes, mixed with boiling water, and baked in the ordinary way.

FRUIT-BREAD.—Stewed apples, pears, peaches, pitted cherries, black currants, or berries, may be mixed with unbolted flour, and made into fruit-bread. A little sugar added will convert the article into fruit-cake.

Rice Apple Pudding without Milk."—Boil rice till nearly done, then stir in sliced tart apples, and cook about twenty minutes.

Griddle-Cakes.—Oat-meal, wheat-meal, or corn-meal, may be made into a batter by mixing with cold water and baking on a soap-stone griddle. Some prefer hot water for oat-meal.

Squash Cakes.—Mix flour, or meal, with half its bulk of stewed squash, or West Indian Pumpkin; add milk sufficient to make a batter, and cook on a griddle.

Soups.—Split-peas, beans, barley, and rice, are employed in the preparation of hydropathic soups. One pint of split-peas, boiled for three hours, in three quarts of water, makes one of the best soups for vegetarians. Some add a trifle of sugar. Bean soup is made of similar proportions, and then boiled in a covered pan for four or five hours. Rice should be boiled until entirely soft. Barley should be soaked for several hours, and then boiled slowly, in a covered pan, for four or five hours. Tomato soup, made in the following manner, is a pleasant and wholesome dish: Scald and peel good ripe tomatoes; stew them one hour, and strain through a coarse sieve; stir in a little wheaten flour, to give it body, and brown sugar in the proportion of a tea-spoonful to a quart of soup; boil five minutes. Okra, or gumbo, is a good addition to this and other soups.

Food of Vampires.

There is a Spanish fable, founded on a very ancient fact, traditionally descended from the earliest population of that country, to the effect that persons buried in a *trance* would, in a few weeks, revive, and ask for food and fresh air. On one occasion, the learned doctor ordered the arisen patient to drink a cup of blood taken from the arm of a healthy person. The result was astounding. While the man, from whose arm the blood was drawn, soon languished and died (no doubt from the operation of some other cause,) the patient, who so greedily imbibed the crimsoned fluid, rapidly recovered, and lived long in the land. But he was feared by the superstitious, and after his death, the most alarming stories were told of his midnight wanderings in quest of living blood.

In a few instances thereafter, where persons, entombed in a trance state, experienced a bodily resurrection, the doctors fed them the blood of bullocks. This, also, had the restorative influence, and so rapidly, that *blood puddings* became a rare but popular dish, in Southern and Central Europe.

This was the origin of the fable of Vampires. But it would be interesting to trace the fancies of mankind upon the Vampirism of the middle ages, and even not more than an hundred years ago. See the very extraordinary report of a surgical and military commission, made about a century since, on the subject of Vampirism in Hungary, or some province adjacent—many of the victims were " hey-dukes." The so-called Vampires, it would appear, were buried alive, in a state of epidemic trance, and their neighbors complained of being grievously haunted by their appearances. A number of them were exhumed (some after a burial of many weeks,) and exhibited signs of life, in fresh blood at the lips, and in cries and groans, when their heads were cut off, or when a stake was driven through them, as they revived.

Naturalists, rejecting the demoniac significance of the term, apply the word to various kinds of bats and vulturous birds; many of which, even by the best writers, are supposed to destroy human beings by sucking their blood. But this is no more true of certain bats, than it is of spiders and other crawling creatures of the tropics.

There is a moral, however, to all this: that human beings

should never be entombed alive, nor while in a death-like trance, however profound, but only *when real disorganization shall have commenced.*

The Uses of Animal Food.

In contemplating the future of humanity, nothing is more vividly clear than the total abolition of all animal diet. Men will not slay and eat when they become true philosophers and spiritualized poets. But the world is physiologically adapted, at the present era, for the consumption of animal substances. Any total abstinence will be attended with an exaltation of the mind, a quietude of the pneumogastric and sympathetic nerves, an indisposition to hastiness of temper, and an increase of bodily powers to resist disease. And yet, we do not recommend total abstinence, but would urge a temperate use of fish, flesh, and fowl—*i. e.*, not more than once per diem, at dinner. And it would be far better to abandon their use utterly during four months of every year. Select the months when your appetite dictates a change. We suggest April, May, July, and August.

Beefsteak, when eaten to excess, *will swear outright* in human speech and conduct. But it is very remarkable that *a fruit and vegetable diet*, persevered in for years, both winter and summer, *will not build factories and elaborate railroads.* If we insist upon the development of gigantic enterprises, the diets of the gods must be suitable for giants. Poets, musicians, artists, and the truly spiritual, will not so construct and execute; therefore, they need not the ferocious and heroic diet. Food, like dress, must have reference to occupation. A frail physical system, with enfeebled blood-making powers, will survive longer, and accomplish more, on a moderate meat diet.

Food as a Medicine.

The human system is pervaded by all its impressible elements. Every organ in the brain is a center whence irradiates an influence, which, blending with the emanations perpetual from each neighboring center, mingles and diffuses itself through all subordinate parts. Hence, either food or medicine exerts an influence upon the whole constitution at the moment of absolute contact. All good medicines are, therefore, nutritious to the entire system, and all good food is, or may be, medicinal in its influence—as water, for example, which, at the same moment, is both nutritious and therapeutical.

The best *Materia Medica*, therefore, is the best apportioned cupboard of food; and the judicious use of eatable and drinkable substances, is the best medical treatment. Certain articles of food will bring on constipation of the small intestines; consequently, those substances will either cure or palliate undue relaxations, diarrhœa, and watery discharges. Some kinds of food will generate mucous, filling the healthy system with *catarrhal* predispositions, including lung and throat irritations and oppressiveness; therefore, those substances—butter, beef, eggs, oysters, and wild game—may be depended upon, in moderate quantities, for all the blood and vital strength which the debilitated may require.

The Hahnemannian School has fully demonstrated the instantaneous and universal effect of a medicine upon the entire constitution on the moment of proximity, and by so doing, our philosophy is established, which teaches that there can be no boundaries fixed between medicines and articles of food. Human beings get sick under the influence of food. Does not this prove the power of food to produce health? Physicians are *needed* more to instruct the world regarding the rules of common sense in human life. Heat, light, air, exercise, sleep, food, clothing, magnetism, water—behold! these are our great good Physicians. Where these doctors prevail, the people are all healthy, happy, and beautiful.

The Evils of Gluttony.

It has been the remark of every intelligent foreigner who has enjoyed the opportunity of studying our social customs, that no people on the globe eat as much as Americans. Medical men in our midst attribute the prevalence of dyspepsia chiefly, if not altogether, to this gluttony. We begin, they say, to stuff in childhood, we continue it through adolescence, we persevere till middle age; and when, at last, the stomach succumbs, worn out by over-tasking, we lament for the rest of our lives our compulsory abstinence. It must be confessed that these charges are substantially true. In the dwellings of the comparatively poor, as well as in the households of the rich, there is, on the average, twenty per cent. more food consumed than nature requires, or health will allow. So universal is this national foible, that few are really aware of it. Hundreds and thousands of men, women, and children, in these United States, daily eat to excess, yet will smile incredulously when first accused of it.

To this gluttony, more than our climate, is to be attributed the sallow complexions we have as a race. To this, also, may be traced that tendency to depression of spirits which is equally characteristic of us. The over-loaded stomach is never permitted to have rest, and, as a consequence, the skin acts sluggishly, the mind feels dull, and a lassitude pervades the entire system. Let any one of our readers, who has been accustomed to hearty meals, and who complains that he feels stupid after dinner, and indisposed to work, try for a week or two a more frugal diet; and our word for it, he will find that he sleeps better, that his animal spirits are higher, that his intellect acts more clearly and rapidly. Even his personal appearance will improve, his complexion becoming clearer, and his eye brighter. Nor is it our sex alone that would be benefited by more temperate eating. Many a fair belle who now flies to cosmetics, would be vastly lovelier, without the aid of art, if she were less fond of luscious cookery, and more self-denying in disposing of it.

It would be a curious, but instructive calculation, to count up in dollars and cents the amount wasted in the United States, in excessive eating. We have no doubt that the aggregate would buy up food, year by year, for all the famine-struck nations in the world. Millions would be inadequate to represent the total. Think how much greater the realized wealth of the American people would be, if a healthful temperance in eating had characterized them for the last two generations! There would have been, in that event, more than enough to liquidate our whole foreign debt, including all the money borrowed for our railroads, canals, coal mines, &c., &c. It is probable that, on the average, one-fourth of the money now spent on the table, everywhere throughout the Union, might be saved, not only without injury to those eating, but positively to their advantage. Intemperance in food is almost as general as intemperance in drink once was, only the former, from being less immediately deleterious, does not attract so much notice.

The Secret of Longevity.

Perpetual youth (says a modern poet,) was the fountain for which the chivalrous Spaniards sought, with the enthusiasm inspired by sincere faith in its existence. That there is far more youth for the human race than is enjoyed, there is no doubt. The average life of man has been, and ever will be, affected, in its length and pleasures, by his habits. It is, we believe, no fable, the tale of men living centuries in the earlier ages of the world, when the habits and pursuits of men were purer and simpler than they since have been, and now are. There is, even now, a great difference in the longevity of men of different races and nations; a difference traceable to, and only to be accounted for, by the difference of their manner of life. In our communities, we can mark the same difference in the average term of life, by glancing over the different castes and professions of society. The author, artist, lawyer, physician, and merchant, are not proportionately long-lived, compared with the men who dig trenches, carry the hod, wield the sledge, or guide the plow. Why is it? Because life is sooner fretted or worn out by excessive than by slight friction. The man who labors with both body and mind, whose sphere of action gives greater exertion to the nerves, upon which sensibility and the acuteness of the intellect rest—whose physical and mental wheels whirl electrically—cannot, and, if he studied the philosophy of cause and effect, would not, expect to live as long as he whose machinery felt but the equable flow of life's current. The most intellectual, as a class, are proportionately short-lived. They are so by reason of the greater friction to which their organization impels them, and also, because, as a class, they seek and accept artificial stimulants, which, however accelerating to vivacity or power for the moment, fearfully cut short the term of life, and which the man of quieter organization seldom or never indulges in. The American people, could they have their tastes and feelings so changed, that honest, peaceful agriculture, would be the goal of their physico-industrial ambition, and the whole nation be transformed into simple-lived and happy peasantry, with plenty to eat, drink, and wear, and no inordinate craving for more—three generations would not pass without lengthening their average life at least one-fourth. They now hurry themselves out of the world, by exciting and straining the delicate net-work in which soul and body are compounded. And this goes to prove that life, peace, and pleasure, are granted to man just in proportion to the truthfulness of his thoughts, habits, and pursuits.

All professions or acts which accelerate decay and death are more or less false. Byron thought gin enabled him to compose more freely and rapidly than he could otherwise have done. Other authors, artists, and high intellects, have thought the same of coffee, opium, and tobacco, and for the moment were correct; but every unnatural tension of nerve and spirit produced by these agencies brings a reaction, the effect of which is the loss of stamina—of life. Byron, without his gin, would have been not only clearer and cooler headed, but his nerves would have stood by him a quarter of a century longer. He burned them by his intensifying process. The true source of the fullest physical, mental, and moral power, will, by-and-by, be understood as based upon the simple and natural healthfulness of man's organization; and as man learns that his life can be prolonged and made more valuable, pleasurable, and noble, by keeping its fountain healthy, the false pursuits, habits, and agencies, which now stimulate him to disease and death, will be abandoned. *There is no reason why man should not live as long now as in the patriarchal and truly Arcadian days.* Nothing prevents it but his habits of life.

Magnetic Disturbances.

Mankind are very impressible to magnetic disturbances, especially in certain conditions of the atmosphere. The skin is organized so exquisitely, that every such disturbance is likely to ultimate upon it, or else in the glands of the bowels, throat, or lungs. The almost certain preventive of scarlet rash and consequent fever, is, first, bathe the body in cool (not cold,) water; then dry it with the hand; and lastly, anoint the whole surface, and wrap it in soft flannel. No stimulating drinks or heating foods; only rest and nursing. Let the *Self-Healing Energies* operate through a few days of quiet and human magnetism.

REMEDIES.—Unnumbered remedies suggest themselves for human ills, but we do not make mention of them, because, what will operate effectually in New York or Massachusetts, would have little beneficial influence on the same disease in Illinois or Iowa Hence, we give only such general directions as cannot fail in the multitude of cases, and in any latitude. In order to get the full benefit of the *Self-Healing Energies*, and to unfold to your own understanding the philosophy of Nature's curative principles, we admonish you to spend no more money for medicines, but save your dollars, and, when diseased, avail yourself of Rest, simple Food, peaceful Spirit, and the Magnetisms of Friendship.

Antiquity of the Bath.

The delightful and health-sustaining practice of BATHING—(says Culbertson's Circular,)—may be traced to the earliest times. In many of the ancient religions, the purifying effects of water had either typical or virtual significance. The Hindoo temples, as well as those of most other nations, were generally by the side of some stream, or were furnished with artificial means of bathing; while in the Hebrew ritual, cleanliness is specially enjoined. All had their sacred rivers, and travelers are still told of the virtues of the Nile, Ganges, and Jumna. Through the lapse of centuries, from the days of Noah to Presnitz, water has been esteemed a remedial agent. The baths of the Romans were often magnificent, and although despoiled and ruined, the traveler is still amazed at their extent and number. In Russia, the use of the Vapor Bath is general, from the Emperor to the poorest serf. Through all Finland, Lapland, Sweden, Norway, and the vast Northern Empire, there is no but so destitute as not to possess its family Vapor Bath. Equally general is the use of the Vapor Bath in Turkey, Egypt, and Persia, among all classes, from the Pasha to the poorest camel-driver or Arab boatman. Even the Red men of our forest have more fully appreciated the advantages of the bath than their more civilized successors. They had, and still have, their Vapor Baths, by throwing water upon hot stones, and standing or sitting in the steam. In America, the Vapor Bath is much less known and appreciated than in the continents of Europe and Asia, and yet, in no country are they more needed. In England, and all other moist climates, the skin is kept in a supple condition, the cuticle pliant, and the pores, or orifices (every one of which is an excretory organ,) are habitually opened. It is for this reason that the blood is so readily discernible, giving a fine color to the face of the English lady; and thus, we find the English complexion better than that of France and many other countries, while we rarely meet with that pallid appearance so common to the inhabitants of our cities, or the sun-burned and opaque complexion of the inhabitants of the country. While all other organs of the body are equally healthy with those of England, the immediate surface of the skin is less so, and hence, all those diseases common to the surface are more prevalent in America than in England. Some of these diseases of the surface are readily communicated to any weak organ, and thus, in some parts of the country, colds and coughs prevail.

Pseudo Health.

The reader will consider himself hereby authorized and duly commissioned to "whisper" the ensuing statement to every individual who "*enjoys* miserable health," or whose symptoms are variable—now up, now down—under the treatment of some college-bred physician, to wit: *That it is impossible to communicate health and strength to the human body.* Let the remark be a thousand times emphasized and enforced from the mouth of every sacred canon: That no human body can be *stimulated* up into a healthy and harmonious condition. The constitutional vigor is increased or diminished only by and through the

Pneumogastric and Sympathetic Nervous Systems; and these Systems, with all their manifold complications and diversified ramifications, are dependent upon that Power of Will, which, (being warmed by *love* and directed by *wisdom*,) occupies the chambers and convolutions of the brain. All health is born legitimately. It, like Venus, arises out of the "foam of the sea" of life—the BLOOD. Nothing can develop Blood save the process of digestion, the organs of which, in order to fulfill their beautiful mission, require to be supplied with plain, nutritious food. A roasted potato is more nutritious than eight ounces of fish or fowl; and a quart bowl of porridge, made of equal parts of Indian meal and rolled Graham crackers, is more health-generating, and more strength-giving, than two pounds of porterhouse beefsteak. In like manner, you will find it chemically true that five long, deep, well appropriated inspirations of pure air, are more invigorating than a cup of either wine, brandy, or coffee.

Tea, Coffee, Alcohol, and Tobacco.

The following is extracted from a letter written by a young man of this city:

"Is it possible that it became necessary, in the progress of the world, that mankind should degenerate physically, and be exalted, or strengthened, mentally? Is the mental temperament gaining on the nutritive and muscular? Did man need such drugs for any such purpose? In a word, have they a mission, directly or indirectly beneficial to the race? Just as a young duck naturally takes to water, so did the writer love *Tea, Coffee,* and *Tobacco,* before he used either. My experience teaches that most drunkards are men above mediocrity. E. A. Poe is a remarkable instance. And it really seems that the brightest, keenest, most refined and delicately organized minds, are most susceptible, and the readiest victims of stimulus. Why is this? Please reply to   A YOUNG MAN."

EXPLANATION.—When mankind were simple, unartistic, ignorant, and undeveloped, their physical and mental wants were few and simple, and, therefore, easily satisfied. Limited capacities demand limited means of subsistence. The simple-minded have simple desires, and such seek only simple gratifications.

But as the law of Progress began to swell the buds of human intelligence, and to expand and enlarge man's capacities for discovery and mechanical inventions, in that same proportion did the *excitability*, and *irritability*, and *irrascibility* of the mind increase and multiply. High susceptibilities of organization are indispensable to the development of Art, Music, Science, and Spiritual Experience. Poetry, Philosophy, Literature, Architecture, &c., are impossible to coarse and low-toned minds. Exalted refinements of the voluntary parts of the brain, and great delicacy, also, of the whole ganglionic nervous web-work thereunto attached, are prerequisite to the conception of fine principles in any department of human interest.

Parents, of such exquisite refinement and consequent genius, transmitted, not their talents and superior proclivities to their offspring, but, instead, all the *excitability*, and *irritability*, and *restiveness;* and thus, many times, the children of very superior parents receive only such miserable inheritance by the fires of procreation. Consequently, as you readily perceive, such offsprings are diseased in many ways, especially in their nervous systems; and such, therefore, search for the appropriate remedies as "a young duck naturally takes to the water."

Originally, and by themselves considered as medicines, the various stimuli used by different races are pure and appropriate remedies for diseases of the nervous system. Tobacco, coca, opium, hashish, alcohol, tea, coffee, &c., are beautiful and appropriate medicines. And the all-wise Father principle of man's existence unerringly led the East Indian, the Greek, the Italian, the Chinaman, the Monks, the Indian of America, and the nervously-diseased of all countries, to the perfect remedy for the evils absorbed by parentage and from the mother's milk. Hence, it is not strange that many desire these stimulants even before they use either—for diseased organs, covered with their appropriate ganglionic nerves, are instinctively actuated toward the natural means of procuring health. The brightest and most promising persons are sometimes irresistibly moved to subdue the irrascibility of their high-toned organizations. Thus, the most " delicately organized are the readiest victims of stimuli." But the roughest and grossest natures, with low-toned and limited mental capacity, join the army of intemperance from sheer *imitation*, when young ; and subsequently they continue in such habits, even after judgment and conscience rebel, from the tremendous force of a misdirected appetite, which will not " down at the bidding."

Our meaning may be better understood regarding stimuli, if we here remark that, according to our investigations, alcohol and the popular drinks are not irritants or stimulants, except in their primary, or first and immediate, effects upon the ganglia and nervous system. Their secondary and absolute effect is the exact opposite of excitement and irritation. Tobacco, opium, tea, coffee, &c., are debilitants, and prostrating to the nervous organism. They generate a temporary fortitude, beget a stoical indifference, and a disposition to take no part in surrounding vehemence and boisterousness ; they impart a fictitious equanimity, a tranquilness, an imperturbable composure, amid the raging delirium and fuming excitements of a discordant society. And it was solely in consequence of these desirable effects that the human race was, in the first instance, led to the discovery and use of several stimulants. But, strange to say, these same medicines will generate *exactly opposite effects* in the nervous systems of those who, in health, use them as socializing agents and as habitual luxuries ; that is to say—tobacco, tea, coffee, rum, &c., will beget in healthy persons the same *excitability*, and *irritability*, and *irrascibility*, which symptoms these remedies are given by the God of Nature to destroy in "those that are sick."

But the time has at length come on earth, we think, when people may erect the standard of Health. This standard will permit the use of remedies only in cases of disease and suffering. All stimulants and weeds are medicines. If our parents, by the force of any imaginable cause or habit, imparted to our nervous system *a love for stimuli*, it is our individual prerogative to exalt Will, and Wisdom, and Love, above the hereditary bias. "A young man" must be A MAN—on his own account, and for the promotion of his own individual prosperity, both on earth and in the Spirit Land. His dear mother may have lived and worked through long years of suffering and excitement, the friction and hardship of which she may have sought in ignorance to overcome, by recourse to *tea and coffee*, or, perchance, by the use of *opium and snuff;* so, too, his headstrong and not well-educated father—fired with energetic blood, and with the irrascibilities consequent upon some transgressions, either physical or mental, may have resorted to alcohol and other so-called stimuli, as means of oiling the wheels of life— imparting the mad wish (or appetite,) for delightful sensations to the nervous system of our "Young Man." But there is but one absolute remedy, namely: INDIVIDUALIZE YOURSELF, and set out to control your appetites by the WILL, which is the center of gravity in the possession of every individual mind. Remember: The gods help those who help themselves—in other words, you will have plenty of friends when you do not need them. Hence the necessity of SELF-FRIENDSHIP.

Self-Rectification in this World.

STATEMENT.—"In my soul there is a feeling of evil done by itself to itself. Having lost all my former faith in oral confessions, and being without hope of relief through importunate prayers—yet feeling that my nature needs something arbitrary to guide it, something external to do as a penance—I take the liberty to inquire how, in the light of Spiritualism, am I to satisfy my own soul, how reconcile myself to myself, how overcome the evil effects of past deeds, that I may become once again happy in my heart?"

REMEDY.—There is a glimmering of truth in the Catholic doctrine of penance imposed upon sinners as a preparative remedy for the removal of their sins. But the philosophic method, which can heal in any measure the wounds self-inflicted on one's own spirit, by non-obedience to the highest attractions of the soul, is this: To cheerfully and promptly set about the performance of all possible offices of benefit to the universal brotherhood of Man—commencing with self-justice, both physical and mental, which includes the happiness of the other self (the conjugal counterpart,) and extending such kindly offices and offerings, whenever opportunity shall offer, through every link of the golden chain which unites man to man, the human soul to Mother Nature, and all, in one glorious eternal union, to Father God.

The Meaning of the Term "Principle."

In the New Philosophy, the term *Principle* is employed with two significations, which should be observed by every one who sets out to master its metaphysical and trans-mundane teachings.

"Principle," in the first place, means an *immutable mode of action*. In this application of the term, we mean to embrace every expression of matter, also every established and unalterable rule with respect to mind or science. A mathematical *principle*, for example, signifies the unchangeable law that regulates the science of numbers. In like manner, a physiological *principle* means the fixed mode of action natural to organs and functions. When the mind thinks of Principle, with this application of the term, it thinks of an immaterial, non-substantial, undiscernible *rule* or mode of being and acting.

"Principle," in the second place, means an immutable and immortal *substance*—ethereal, spiritualized, beyond the detection of the five senses, yet as real, as material, as much the opposite of nothing as anything *substantial* and indestructible can possibly be.

In this sense we employ the term *Principle*, when speaking of man's soul and spirit; also, when referring to Nature or to Nature's God. "God is a spirit"—the same is true of Nature —and both are substantial, inter-intelligent PRINCIPLES. Their immutable modes of action are also "principles," but this is true only in the first definition : for the second use of the term is applicable to *that which acts*, rather than to the action.

A person may discover and conceive of a *method* of matter, and he may call that method "Principle," and yet it is possible that the same person entertains no clear understanding of *that* which thus acts before his senses. In such case we term the man a sensuous or a scientific reasoner ; but if he can comprehend the second meaning of "Principle," then, we say, he is a metaphysician and a spiritual philosopher.

Which is the First—the Body, or the Spirit?

QUESTION.—Does the spirit grow up out of matter ? Is it the refined product of material organization ? Or, is the material organization the manifestation and product of spirit ?

ANSWER.—All formation and organization are the effects of eternal Principles, which are inter-intelligent and divinely perfect in all their operations. These Principles are : Association, Progression, and Development. The first manifestation is a revelation of the first law ; next comes an unvailing of the second principle; lastly, the third principle is disclosed in all the splendor of its immortal powers. In all parts of the "stupendous whole" these three Laws are fully revealed to the mind of the philosophic observer.

Now comes the answer to your question: Matter and mind, which are two principles, move onward as essentially One substance. The two eternal principles begin at the fountain vortex, flow up from the abysmal depths of the God-center, and, *together*, move outwardly through suns, planets, minerals, vegetables, and animals—terminating their diversified pilgrimage in the formation of the human type. What is true on our earth, is equally true on thousands of other planets in space.

Body and soul grow up simultaneously upon the summit of all the lower kingdoms. Man's *spirit* is unfolded by the joint action of his body and soul. The spirit is not first, but last, in order of development. Do not understand us to teach that body and soul, like husband and wife, become parents to the spirit (in the sense of creating it,) but this: that the SPIRIT is attracted to and enthroned within the body-and-soul temple, by means of the perfect *order* and *form* which the twain establish in the embodiment. Spirit, therefore, becomes sovereign over both body and soul, as it were, by invitation and attraction, and so man's individuality is fixed.

After man's individual existence is established, then the three *actuating* Laws, before mentioned, commence and continue their work among the mind's attributes.

Incorrigibleness of Blood.

QUESTION.—Does a religious education overcome, as some assert, all the inherited predispositions to evil?

ANSWER.—No; for the original or progenitory bias of blood is the root-life of the mind. Education is a *graft* upon the life tree, which will bear fruit like unto itself, but the vital source is *organization*. You will remember Amintor's speech to Evadne, in the play of the "Maid's Tragedy," wherein antenatal propensities, stronger than education, are thus painfully acknowledged:

> ———" Do not mock me
> Though I am tame, and bred up with my wrongs,
> Which are my foster-brothers, I may leap,
> Like a hand-wolf, into my natural wildness,
> And do an outrage. Prithee, do not mock me!"

An important question suggests itself just here, with respect to the *remedy* for constitutional defects of character, but which we will not now consider. Let the first birth be *right;* a treasury of pure love and sweetness; a harmonious infancy "that opens like perpetual spring."

Do Infants grow up in Heaven?

QUESTION.—Do infants grow up to the proper stature after death? If so, shall we recognize our little ones when we go to them?

ANSWER.—There are many things to be written on the subject of infants, some of which may come up hereafter for consideration, but now we will only answer the question propounded.

One who had died an infant, and become an adult in the spheres, returned to its parents yet on earth, and caused the medium psychologically to perceive it as it appeared before death, some twenty-two years previous. The seeress accordingly reported the departed one as yet in its infancy. The description was so complete, in every particular, that the listening parents could not but acknowledge the presence of their darling spirit child.

In like forcible and convincing manner, when parents enter the Spirit Land, the grown-up children gone before will, by certain internal signs and intuitive appeals, perfectly satisfy the parents of the relationship subsisting between them. Not by psychology, but by the force of intuitive wisdom, by which time is annihilated, and the spirit is enabled to comprehend its own history and dependencies. By the wisdom of intuition you will again meet the loved one lost. Remember that Intuition

> "Unvails all Nature's laws and miracles.
> The world's incomprehensible is no more,
> But all is plain as new-born starry spheres."

Why Primary Processes are Discontinued.

QUESTION.—Why are not man and the animals produced now as they were at first?

ANSWER.—Because the ultimate processes of Nature have superseded the fundamental methods; just as the spinning of wool and cotton by machinery takes the place of the hand-loom, the distaff, and the spindle, of the last century; or, to use another figure, just as the human head takes precedence of the body in carrying forward the functions of organic life.

But do not misunderstand our meaning. We know of no era in the earth's history when men and animals were produced in a manner differing from the present universal law of procreation. We, however, do observe a period in the extreme past when plants merged into animal organizations, and another more recent epoch when the highest animal structures started, through the reproductive organism of the prepared female, the lowest types of the human family. But we do not behold any time when the forms of plants became animals, nor any transformations by which the forms of the highest animals became human beings. We teach, therefore, that the human race was started in the *perfected spermatozoa* of the highest animals, but not that the structures of the highest animals were gradually modified and promoted by progression to the form of man.

The reason why primary processes are not continued must be obvious to every thinker.

Unity of the Universe.

The atmosphere of truth is never darkened by the clouds of fear. Only those go astray who do not realize the holy influence of truth. Can you not attract higher intelligences? The chain of causation is never severed, though it may *seem* to be broken, when some unseen Spiritual confusion disturbs the general harmony. Nature's empire is governed by immutable laws. Hence the poet was misinspired when he wrote—

> "In Nature's chain, whatever link you strike,
> Tenth or ten thousandth, breaks the chain alike."

Not so, reader! Nature is the conjugal MATE of the Infinite Mind. Each is a perfect *translation* of the other. And the higher the intellect, and the purer the affections of man, the greater is his appreciation and enjoyment of the "stupendous whole."

Temporary Marriages.

Should one who believes that his soul-companion is in the Spirit World, ever marry in this life? and would not marriage be rendered unhappy by the entertainment of such a belief?

ANSWER.—The divine law of true harmonial marriage—that is, of one man to one woman, each being the other's counterpart and equal—is eternal in its power over human destiny. Only the wandering, the uncompanionated, the pure-loving, can feel at liberty to marry. No heart can yearn in purity for conjugal fellowship, unless its fountain of love is unsought, or if sought, unsatisfied and unmated.

It is impossible, we think, for those who sincerely and devoutly believe that the heart-mate is waiting for them beyond the tomb, to enter upon a worldly marriage—to consent to a mere legal mockery of the genuine—for sensual gratification and temporary purposes. Disappointment and loneliness would infest their earthly hearts, and deadly discords would prowl through their haunts and habitations. Happiness can bloom immortally only upon the life-tree of an eternal conjugal union. But the belief that *no soul is to be forever alone*—that each heart will one day be indissolubly joined to its true mate—should inspire the world with unspeakable joy.

What is Evil?

Many minds are greatly troubled with the puzzle of Christendom, namely, the Origin of Evil.

We answer briefly, that the Origin of Evil is Ignorance, and that the Origin of the Devil is Evil. The blue heavens are a mixture of darkness and white light. Nature is a vast magnetic machine or battery, with a positive and a negative pole, and man is the armature. Man is an intermediate being, and connects the two opposite poles. He joins the animal and the angel worlds into one. The consequence of which is, that he receives the antagonisms of the one, and feels the attractions of the other. Evil is the raw material of this life; the incident of that good which is to come. Darkness is driven away only by the approach of light. Our want of development is the only absolute evil. The Devil never lives in the presence of Wisdom and Integrity.

Where goeth the Soul of Things.

PROPOSITION.—A correspondent writes: "You establish that animals and plants are not immortal, but what becomes of their thinking, sensitive, and life principles after their death? Where goeth their soul?"

ANSWER.—According to our investigation, now continued

for more than twelve years, the soul of nothing is necessarily immortal as to its *shape* or form; neither, in an essential sense, is anything susceptible of annihilation. Only ultimate forms are eternal. The internal form of man is an ultimate form, therefore, his internal form, with its glorious endowments, is eternal. But the forms of plants and animals are transitional forms; therefore, such forms, with all their life principles, eventually cease from existence. Only the form ceases to exist; the life principles eternally continue.

But whither go the soul principles of animals when they die? Into the illimitable sea of life that throbs and surges around, above, and through us. The bodies of animals and plants decompose and blend with the substances of the material world; so the souls of these transient bodies dissolve and mix with the vital principles of the spiritual universe, which encompass us on every side boundlessly. The earth is incessantly producing new bodies, new fish, new plants, new animals, &c., and into these last developments the elementary principles of previous bodies freely enter, on their way up to MAN.

Habits and their Consequences.

QUESTION.—"Although I have read much, I am not satisfied with what I know of *Tobacco, Alcohol,* and *Syphilis.* Why were men so constituted as to love the first two, or engender the last?"

ANSWER.—By a close and cautious examination, we are forced to the conclusion that no human being is constituted to love either alcohol or tobacco. The effects that follow the first use of alcohol are pleasing and tranquilizing. The memory of this temporary relief from worldly annoyances and fatigue, is the magnet that draws thousands to drink more, and more, and yet more, until the habit is fixed beyond the power of will to resist. And this is true even after judgment and conscience have combined against the habit. The use of tobacco is at first an affair of *imitation* among boys; subsequently it becomes an independent, injurious habit, uncontrollable by the individual's will.

Mother Nature is very kind and just to all her loved children. If they obey her best laws, she crowns their deeds with happiness. If, however, they remain under the guidance of lower laws, she metes out to them the philosophical consequences of their misdirection. She crowns conjugal love and true marriage with the diadem of happiness; but she as justly generates disease and corruption in the vitals of those who violate her sanctities.

Put on the Will Power.

DISEASE.—"I am a sufferer from no apparent disease, but possess little *vigor* of either body or brain. I desire to know if this arises from some defect of organization, or is it a weakness I may some time surmount?"

REMEDY.—It is unworthy an immortal spirit to surrender its powers to that lawless wretch, the meanest sinner among five hundred diseases, known in these parts as "GEN. DE BILITY." Concerning the duration of bodily infirmities, we give answer in the poet's words:

> "Naught eternal is
> But that which is of God. All pain and woe
> Are therefore finite. Can the robber steal
> From God? All souls are His, and Him deriv'd,
> And thus are good, and *Good alone is endless.*
> But Evil, having birth from second causes,
> Created things, gross matter, and their laws,
> Is not from all eternity with God,
> But hath a *recent origin*, and thus
> Hath not an endless, but a casual being,
> And must expire where its reign began."

Obey the laws of bathing in cold water every morning; arise

early; go out into Mother Nature's auroral embrace; rest half an hour before dinner; put on the Will Power; step strongly; and thus summon forth your yet dormant energies. You may conquer.

The Evil of Suicide.

Persons who die by sudden accidents do not suffer intellectually and morally, because their misfortune is purely physiological or physical (the same in this respect as that of the suicide,) while the deliberate or desperate taking of one's own life is attended, in the other world, with the superaddition of all the intellectual disqualifications and moral darknesses which were antecedent to, and consequent upon, the state that led to the ignoble act. No individual can go self-sent and abruptly into the Spirit Land, without ultimately discovering the secret cause of the deed to have existed within his own mental constitution. This is certain, no matter how great the earthly provocation which primarily induced the person to consummate the act. It is the vivid consciousness of this individual unworthiness that, for a lengthened period in the other life, causes the regrets and sufferings of the suicide.

Hahnemann's Materia Medica.

There is much truth in the following remarks by Dr. Hempel: Who would not rather give his child a few pellets of hepar sulphuris, spongia, bichromate of potash, etc., to have it cured of croup, than to have it bled, purged, crammed full of emetics, and to have its skin blistered by vesicatories? Or who would not rather be cured of pleuritis or pneumonia in two or three days, by taking a few pleasant, harmless pellets of aconite, bryonia, belladonna, etc., than by submitting to all the tortures of the regular treatment for three times the length of time required by the Homeopathic practice? The obstinacy and blindness of the pretended "regular" physicians are truly wonderful; they all have it in their power to shorten the sufferings of their patients by treating them Homeopathically, and yet, there is not one of them who would only condescend to look into Hahnemann's Materia Medica, much less to call upon those who have studied it, and who would be glad to inform their dissenting brother practitioners of the treatment to be pursued in particular cases, and to gradually initiate them in the principles of our art. There seems to be a gulf between the Homeopathic and Allopathic practitioners, which can never be filled; there can be no intellectual fellowship between those two classes of men, and nothing is left to the Homeopathic physicians except to appeal to the common sense of the public, and the brilliant results of their practice.

An Infallible Catholicon.

The *Eclectic Medical Review* gives the following, from a young lady who was fashionably educated at boarding schools, and indulged in idleness at home, so that there was neither strength nor elasticity in her frame: "I used to be so feeble that I could not even lift a broom, and the least physical exertion would make me ill for a week. Looking one day at the Irish girls, and noticing their healthy, robust appearance, I determined to make a new trial, and see if I could not bring the roses to my cheeks, and rid myself of the dreadful lassitude that oppressed me. One sweeping day I went bravely to work, cleaning thoroughly the parlors, three chambers, the front stairs and hall, after which I lay down and rested until noon, when I arose and ate a heartier meal than for many a day. Since that time I have occupied some portion of every day in active domestic labor, and not only are all my friends congratulating me upon my improved appearance, but in my whole being—mind, body, and spirit—do I experience a wondrous vigor, to which I have hitherto been a stranger. Young ladies, try my catholicon."

How to Live One Hundred Years.

Ralph Farnham, the veteran of Bunker Hill, says : " Though I am in my 105th year, I am not past all usefulness ; I split my own kindling wood, and build my own fires. I am the first one up in the morning, and the first one in bed at night. I never sleep or lie down in the day time, but rise at five, and retire at seven, and this I continue summer and winter. I have always been temperate, and for over thirty years past I have not tasted a drop of spirituous liquors, or even cider. I was never sick in my life, so as to require the attendance of a physician. About twenty-five years ago I broke my thigh by falling on the ice, and had a surgeon to set it, but this is the only time a doctor ever attended me. I live on plain farmers' diet, drink tea and coffee, and eat a very light supper, never eating meat after dinner. I have no doubt it is owing to these abstemious and regular habits, *and the avoidance of medicine at all little ailments*, that my life has been so prolonged. I voted for Gen. Washington for President, and have voted at every Presidential election since."

A Word to Mothers.

Fanny Fern has contributed the following excellent medical instruction to mothers, on the subject of air and exercise : " Consider it your religious duty to take out-door exercise without fail, each day. Sweeping and trotting round the house will not take its place : the exhilaration of open air and change of scene are absolutely necessary. Oh, I know all about ' Lucy's gown that is not finished,' and 'Tommy's jacket,' and even '*his*' coat, *his* buttonless coat thrown in your lap, as if to add the last ounce to the camel's back; still I say—up—and out! Is it not more important that your children in their tender years should not be left motherless? and that they should not be born to that feeble constitution of body which will blight every early blessing?' Let buttons and strings go; you will take hold of them with more vigor and patience when you do return, bright and refreshed, and if every stitch is not finished at just such a moment (and it is discouraging not to be able to systematize in your labor, even with your best efforts,) still remember that 'she who hath done what she could,' is entitled to no mean praise. Your husband is undoubtedly 'the best of men;' though there are malicious people who might answer, That is not saying much for him! Still, he would never, to the end of time, dream what you were dying of: so accept my advice, and take the matter speedily in hand yourself."

The Use of Calomel.

There are numerous and various reasons why we do not approve of the administration of calomel in any case. The best medicine that can be given for an inactive liver, is a thorough mechanical manipulation of the stomach and bowels, including the liver side, and the lower extremities, twice every day, in connection with a judicious use of food, air, exercise, and sleep. In some cases, the administration of mercury, as a medicine, may not leave any serious effects behind, because some constitutions are strong enough to repel both its disease and the scientific nostrums; but in every case where the system of the patient is prostrated by disease, and where it cannot master the mercurial preparations of science, we have observed a subjection of the patient's blood and bone (and brain sometimes,) to the arbitrary action of the physical elements. This is slavery, with calomel for a master.

How to Keep the Teeth Clean.

In order to keep the luxury of a sweet breath, we advise, first, a cheerful and grateful disposition; second, regular meals, temperance in quantity, and a prompt digestion. On the subject of teeth cleaning, we quote the teachings of modern chemistry: Microscopical examinations have been made of the matter deposited on the teeth and gums of more than forty individuals, selected from all classes of society, in every variety of bodily condition; and in nearly every case, animal and vegetable parasites, in great numbers, have been discovered. Of the animal parasites, there are three or four species, and of the vegetable, one or two. In fact, the only persons whose mouths were found to be completely free from them, cleansed their teeth four times daily, using soap once. One or two of these also passed a thread between the teeth, to cleanse them more effectually. In all cases, the number of the parasites are greater in proportion to the neglect of cleanliness. The effect of the application of various agents was also noticed. Tobacco smoke and juice did not impair their vitality in the least. The same was also true of the chlorine tooth-wash, of pulverized bark, of soda, ammonia, and various other popular detergents. *The application of soap, however, appeared to destroy them instantly.* We may hence infer that this is the best and most proper specific for cleansing the teeth. In all cases where it has been tried, it receives unqualified commendation. It may also be proper to add, *that none but purest white soap, free from all discoloration, should be used.*

Sweet Oil in Relation to Poison.

In the first of the Harmonial series of volumes, we urged the immediate use of sweet oil, to be followed by an emetic, in cases where active and deadly poisons had been swallowed. The prescription was treated as worthless by many so-called scientific physicians. Now, however, we learn by the *Tribune* that "M. Blandlet, of France, has called attention to a very curious toxicological fact, namely, that greasy matters have the power of diminishing considerably the solubility of arsenious acid, either in pure water or in acid and alkaline liquids. Thus, in contact with grease, the poisonous properties of arsenious acid are much decreased, and, at the same time, it becomes more difficult to render its presence evident by chemical reactions. A very slight quantity of greasy matter, according to M. Blandlet's experiments, reduces the solubility of arsenious acid to one-fifteenth or one-twentieth of what it is when it is in a pure state.

"*This explains why arsenic*, taken in the form of powder, *remains sometimes for a considerable interval in the body without producing injury;* it explains also how it is that, in cases of poisoning by arsenic, this substance has not readily been detected in such portions of the body or the aliments which contain much grease. It seems to teach, also, that cream, for instance, is an excellent antidote for arsenious acid. Morgagni states, in his writings, that, in his time, the Indian boatmen used to astonish the bystanders by swallowing, without hurt, large pinches of arsenious acid, having taken the precaution beforehand of drinking a quantity of milk, or eating some greasy matter. As soon as the public retired, the boatmen got rid of the poison by vomiting."

Change of Clothing.

Do not be in haste to put off woolen garments in spring. Many a "bad cold" (who ever saw a good one?) rheumatism, lumbago, and other aches and pains, are lurking in the first sunshiny days, ready to pounce upon the incautious victims who have laid aside their defensive armor of flannel. All *sudden* changes in the system are attended with more or less of danger, but the body can accommodate itself to almost any condition, provided it be assumed gradually. The use of flannel guards against sudden changes of temperature. In a warm day, when perspiration flows freely, if it be allowed to pass off rapidly, the quick evaporation carries with it much heat from the body, and a chill may be produced, followed by the derangement of some function: as "cold in the head," or unnatural discharge from the bowels. Flannel contains much air in its meshes, and is, therefore, a slow conductor of cold or heat. Evaporation proceeds from it more slowly than from cotton or linen, hence its excellence as a fabric for clothing Many persons wear it next to the skin the year round, and find it a shield against prevalent complaints in summer. No general rule can be given as to this; it must depend upon the constitution and employment of the individual. In all cases, however, flannel should not be laid aside until the weather is settled permanently warm.—*Med. Journal.*

An Unhealthy Occupation.

Nothing is more beneath the estate of either Man or Woman than sacrificing *health* because some money is to be gained by it. If your occupation is not agreeable to your happiness and bodily health, *leave it at once and forever*, just as you should flee from the presence of any o'er-mastering evil, and henceforth pursue some calling in which you will feel at home and healthy. Allowing the love of dollars to eclipse our manhood, and especially to do evil that a selfish good may accrue, is worthy the execration and the condemnation of every honest and philanthropic mind.

Impure Water.

Set a pitcher of water in a room, and in a few hours it will have absorbed nearly all the respired and perspired gases in the room, the air of which will have become purer, but the water utterly filthy. The colder the water is, the greater its capacity to contain these gases. At ordinary temperatures, a pint of water will contain a pint of carbonic acid gas, and several pints of ammonia. The capacity is nearly doubled by reducing the temperature of the water to that of ice. Hence, water kept in a room awhile, is always unfit to drink. For the same reason, the water in a pump-stock should all be pumped out in the morning before any is used. Impure water is more injurious to the health than impure air.—*Ec. Med. Journal.*

The Arsenic Eaters of Styria.

According to an article in the *Pharmaceutical Journal*, arsenic is commonly taken by the peasants in Styria, the Tyrol, and the Satzkammergut, principally by huntsmen and woodcutters, to improve their mind and prevent fatigue. The arsenic is taken pure, in some warm liquid, as coffee, fasting, beginning with a bit the size of a pin's head, and increasing to that of a pea. The complexion and general appearance are much improved, and the parties using it seldom look so old as they really are. The first dose is always followed by slight symptoms of poisoning, such as burning pain in the stomach, and sickness, but not very severe. Once begun, it can only be left off by very gradually diminishing the daily dose, as a sudden cessation causes sickness, burning pains in the stomach, and other symptoms of poisoning, very speedily followed by death As a rule, arsenic eaters are very long lived, and are peculiarly exempt from infectious diseases, fevers, &c.; but unless they gradually give up the practice, invariably die suddenly at last. In some arsenic works near Salzburg, the only men who can stand the work any time are those who swallow daily doses of arsenic—the fumes, &c., soon killing the others.

Treatment of Frost-bitten Persons.

A medical writer says: "With regard to the treatment of frost-bitten persons, the part affected should be rubbed with cold water or snow, and then with fluids of a medium temperature, in a cold room; cautiously bring the patient into a warm atmosphere, and administer small quantities of cordials or warm tea, then cover him up in bed, and encourage perspiration. Even where the patient seems quite dead, or has lain as if dead for days, you must give a fair trial to these remedies. When poor Boutillat, the French peasant, who awoke crying out for drink after his four days' sleep in the snow, was brought to his friends, they wrapped him in *warm* linen, dipped in aromatic water, and this was but too probably the cause of the poor fellow's feet mortifying.

How cold slaughters its victims we do not exactly know some say it paralyzes the heart; others think that the cold, to use a popular expression, drives the blood inwards, and kills by apoplexy. The irresistible sleepiness that creeps over a person 'lost in the snow,' is well known, and has often been described; if once it is yielded to, death, under the forlorn circumstances usually present, is sure to result. But, undoubtedly, it may kill at once. Persons have been found stone dead standing upright at their posts, all the machinery of life having stopped at once—the mouth half open, as it was when the last groan was uttered; the limbs still in the position they assumed during life, and having undergone, through the peculiar antiseptic nature of the cold, none of the changes we find after other forms of death."

Influence of Cod-liver Oil and Cocoa-nut Oil.

Dr. Thompson, in a paper read before the Royal Society, states, that he found that, during the administration of cod-liver oil to phthisical patients, their blood grew richer in red corpuscles. The use of almond oil and of olive oil was not followed by any remedial effect; but from cocoa-nut oil results were obtained almost as decided as from the oil of the liver of the cod. The oil in question was a pure cocoa oleine, obtained by pressure from crude cocoa-nut oil, as expressed in Ceylon and the Malabar Coast, from the dried cocoa-nut kernel, and refined by being treated with an alkali, and then repeatedly washed with distilled water. It burns with a faint blue flame, showing a comparatively small proportion of carbon, and is undrying The whole quantity of blood abstracted, for analysis, having been weighed, the coagulum was drained on bibulous paper for four or five hours; weighed, and divided into two portions. One portion was weighed, and then dried in a water-oven, to determine the water. The other was macerated in cold water until it became colorless, then moderately dried, and digested with ether and alcohol, to remove fat, and finally dried completely, and weighed as fibrin. From the respective weights of the fibrin, and the dry clot, that of the corpuscles was calculated.

The Brain of the Negro.

According to the investigations made by Prof. Tidemann, of Heidelberg, the weight of the brain of an adult male European varies from three pounds and three ounces to four pounds and eleven ounces troy weight; that of the female weighs, on an average, from four to eight ounces less than that of the male. The brain usually attains its full dimensions at the age of seven or eight, and decreases in size in old age. At the time of birth, the brain bears a larger proportion to the size of the body than at any subsequent period of life, being then as one-sixth of the total weight; at two years of age it is one-fourteenth; at three, one-eighteenth; at fifteen, one-twenty-fourth; and in the adult period, that is, from the age of twenty to that of seventy, it is generally within the limits of one-thirty-fifth and one-forty-fifth.

No perceptible difference exists (says Prof. Tidemann,) either in the average weight or the average size of the brain of the negro and of the European, and the nerves are not larger, relatively to the size of the brain, in the former than in the latter. In the external form of the brain of the negro a very slight difference only can be traced from that of the European; but there is absolutely no difference whatever in its internal structure, nor does the negro brain exhibit any greater resemblance to that of the orang-outang than the brain of the European, except, perhaps, in the more symmetrical disposition of its convolutions. The facial angle in the negro is smaller than in the European.

Effects of Tea on the Body.

The general theory of chemists hitherto has been that tea lessens the waste of the body, and so sustains the bodily powers with less nourishment than is otherwise required. Dr. E. Smith, at a recent meeting of the Society of Arts, gave the result of some experiments he had made to ascertain the truth of this theory. He found that if there was abundance of food in the system, and that, especially, of the farinaceous or fat kinds, tea is a powerful digestive agent, and, by promoting the transformation of food, it aids in nourishing the body; but with a deficiency of food, it wastes the tissues of the body, and lowers the vital powers.

The Evil of Over-dosing.

"If a Christian man offers a proper prayer to the Lord, short and to the point, and then stops, as all sensible men do, many persons think *it's no prayer at all, and that the petitioner must be a very stupid sort of a fellow, not having sense enough to know how to pray.* Such persons would read Christ a lecture, if he were here now."

The foregoing sensible remarks are taken from the "Union Baptist Church" organ, the *Christian Banner*, published at Fredericksburgh, Va. We respond, "Amen," and proceed further to remark, that most persons think that a doctor who does not prescribe an *enormous quantity of physic* to a sick patient, is no doctor at all. "Daughter, thy faith hath made thee whole," is too Homeopathic for the great majority of mankind. They dose the Almighty with large quantities of pray r, for the same reason that they swallow large doses of medici e, namely, *because they have not progressed out of ignoran c, superstition, and error.*

How to make Iron Magnets.

QUESTION.—"Is it possible to bring common cast iron into the state of permanent magnetism? Is such magnetism better than the hand in disease?"

ANSWER.—It is well known that any bit of iron, properly related to the earth's magnetic currents, will become perfectly magnetized. According to the Professor of Natural Philosophy at Louvain, the artificial process of magnetizing cast iron is simple. Cofford Crahoz, at a late meeting of the Academy of Science, at Brussels, presented the Professor's *modus operandi*, which is equally practicable for any person, who, with economy, wants a permanent magnet. The Professor takes a bar of gray cast iron, which he brings to a red heat, and then sprinkles it on both sides of three-fourths of its whole length with *prussiate of potassia*, and puts it, then, in cold water, and when cooled, it is magnetized with a strong electro-magnet. It may also be used for a horse-shoe magnet. The human system is composed upon magnetic principles, and is, therefore, the best battery for healing purposes.

Food for Infants.

Says a writer: No mother would feed a child on the milk of a sick cow, if she knew it; but is there any reason to suppose that *the milk of a sick woman is more healthy than that of a sick cow?* Either must inevitably be sources of disease. The cows in New York, fed on distillery slops, are no worse off, and no more diseased, than thousands of mothers, who live on unhealthy flesh, and drink, not the slops, but the liquor of the distillery, with the additional poisons of tea, coffee, tobacco, and various drug medicines. What with diseased mothers and distillery cows, our children have a hard time of it; and so, ten or twelve thousand die every year in this single city. And this appalling mortality, far more frightful than the cholera, goes on year after year, and nothing is done, because we think it inevitable, and have got hardened to it. I have written upon this subject for years, and I am determined that people shall think upon it. When they have once thought, there is no fear but they will act. There is no man with a human heart in his bosom, and there can be no woman, who must not feel interested in ascertaining the causes of infant mortality, and the means of staying its terrific progress.

American Improvements in Surgery.

Probably no diseases (says the *Illustrated News,*) have been so ineffectively and unsatisfactorily treated as those attacking the spine, the hip, and the knee, the reason being found in the fact that but little has been accurately known of the real nature, progress, and cause of these maladies. A recent discovery has, however, thrown considerable light on these subjects, and promises to be of considerable use in the treatment of a large class of sufferers. Dr. H. G. Davis, a physician of this city, is the author of this discovery. From researches and investigations prosecuted during many years, he has found that wherever any joint whatever is stiffened from any cause, and the ends of the bones are closely pressed together by muscular action, a very small area of each articular surface receives the whole pressure, and by the action of a well known law, this pressure produces waste, first destroying the cartilage, and then attacking the head of the bones beneath. The area of pressure being thus gradually enlarged, the destruction of the bone spreads, until at length the case becomes desperate. The point originally decayed has extended, and spread, and deepened, and now an operation is indispensable, the limb being, perhaps, inevitably sacrificed, and the life of the patient placed in jeopardy. Now it is evident that as, according to these statements, the whole mischief results from the continued muscular pressure on the joint, the first and chief object must be to remove this pressure, and to relax, for a considerable period, the muscles producing it. The simple apparatus which Dr. Davis has devised for this purpose, has been exhibited before the New York Academy of Medicine and the Medico-Chirurgical College. We learn that it has also been adopted by Dr. Buck in the New York Hospital, and with more or less modification by Dr. Sayre and other eminent surgeons. Its results are stated to be very satisfactory.

Bite of a Rattlesnake.

The Petersburg *Express* publishes the following from a reliable correspondent: A carpenter, while engaged a few days ago in pulling down an old house and removing some of the rotten timber near the ground, was bitten by a rattlesnake. In a few moments his finger was swollen to four times its natural size, and a red streak commenced running up his hand and wrist. A deadly languor came upon him, and his vision grew dim, clearly indicating that the subtile poison that was coursing through his veins was rapidly approaching the citadel of life. But a remedy was tried, merely by way of experiment, which, to the surprise of all present, acted like a charm, the component parts of which were *onion, tobacco, and salt*, of equal parts, made into a poultice; and at the same time a cord was bound tightly around the wrist. In two hours afterward he was so far recovered as to be able to resume his work. I knew an old negro who cured a boy who had been bitten by a mad dog, by the same application.

Modern Luxuries Everywhere.

" When we reflect upon the extreme luxury in which the poor and middling classes live, now-a-days, as compared with much richer people of former centuries, we cannot but wonder how our respected ancestors could have existed at all. We have but to go to a closet and turn a faucet, and we have fresh, cold water *ad libitum*. We have but to touch a brass key in the corner, and the gas furnishes us with a fine and plentiful light. We communicate with our distant friends in the twinkling of an eye, or go to see them with the speed of the wind. Yet, all these luxuries, and many more, inexpensive and common to us, were unknown to our forefathers. Who can say the world does not progress?

Electricity in Vital Processes.

May not that great binding chain of the universe, that universal power, that wonder-working principle, whose intensity continues the same at all accessible distances from the earth's surface—"electricity"—be also the origin and universal cause of vitality and life, both animal and vegetable, by which the instantaneous action of thought and feeling is telegraphed throughout the animal frame? Let us inquire, and by way of illustration we will take an acorn or an egg. Now, it is well known that neither an acorn nor any other seed will germinate if kept dry, nor will an egg produce at the common temperature of the atmosphere (at least in this country,) but both will inevitably perish if their position is not changed.

If the acorn, or a grain of wheat, or of any other plant, be buried in *moist* earth, all the requisite conditions necessary to its growth are fulfilled, because we surround the seed with the means from whence the nutriment for the organization and construction of the plant is derived; and the electric circuit being also completed by that simple act, such nourishment is distributed by the circulating current generated, as has just been pointed out; and this electro-chemical process constitutes, in fact, the only vitality of plants. The suspended vitality of seeds may be regarded as analogous to the broken galvanic, or electro-telegraphic circuit, in which the electrical action is suspended.

Positive and Negative.

QUESTION.—The terms Positive and Negative are often used in our Philosophy as distinct and absolute. Will you give your views of the distinctive significance of these terms?

ANSWER.—We have used these terms with two meanings. First, which is an inferior use, to designate the difference between *power* and *weakness*, or between that which is passive, (negative,) and the opposite term for whatsoever is *active* and energetical. But we have a second and *higher* definition, which is most common in the New Philosophy, namely: "Positive" is applied to any power which works from the surface toward the center ; while, of course, " Negative " would signify any equal power or fellow-principle, which commences at the *inmost* and works outwardly to the circumference. With the latter sense the terms are applied to Love and Wisdom—to woman and man.

How the Roots of Plants Feed.

The following query is important : " Can the roots of plants take up only such substances as are dissolved in the ground, and thus prepared for them, or can they themselves dissolve them ?"

This question has been solved by Liebig, and by experiments made before the Society of Natural Science in Carlsruhe. He has proved that the roots of plants, by giving forth some acid—probably carbonic acid—do dissolve the alkali, ammonia, and phosphorus in the soil. Dr. Schimper showed the meeting, as a further proof of Liebig's doctrine, some pebbles, which evidently had been eaten in by roots of plants. The fact was visible ; the process, however, is not yet clear.

This valuable discovery of the great chemist goes clearly to show us why the rains and floods cannot wash out of the ground the substances forming the food of plants ; on the contrary, we now know that the earth takes from the liquids which touch it, and solidly appropriates substances which the roots of the plants again absorb by their action. In the same way we clearly perceive how plants can draw from the soil substances which are solids, and which are not soluble by water.

Corn-fields and Magnetism.

" Will you give your opinion relative to invalids visiting cornfields, and inhaling as much as possible of their atmosphere ? Also, the best method for throwing off the influence or disease received while magnetizing the sick ?"

Answer.—It is of the utmost importance to invalids, and not less to persons in good health, that the pure air of growing vegetation be absorbed without measure. The exhalations of corn-fields are particularly healthy and nutritious. The atmosphere of many flowers is secretly refining, and is emasculating as well. But the emanations from pine, hemlock, and cedar are far more salutary. Meadows are not at all beneficial to the sick. Better visit some high place, where the light of heaven's effulgent sun shines more than half of the day. Repose there, and breathe in the spirit of health and contentment.

After magnetizing the sick, wash your hands thoroughly, and manipulate each by the other until they are perfectly dry and warm. Another washing would make your exemption certain.

Rain and Electricity.

The scientific world of America has not been willing to see anything favorable in our "Philosophy of the Production and Fall of Rain," published in the *Harmonial Man*, but the "British Association" has bestowed more attention on the subject. Mr. Weekes, of Sandwich, writes as follows to Mr. Rowell, who, at the British Association, suggested the possibility of bringing down rain from the clouds at pleasure : " I have from very early life been an assiduous experimenter with electric kites, atmospheric exploring wires, &c. Now, I beg to assure you that it has several times happened that, when my kite has been raised immediately under a distended, light, fleecy cloud, at a moderate elevation, a free current of sparks has passed from the apparatus during some ten or twelve minutes ; *I have suddenly found myself bedewed with a descent of fine misty rain ; and on looking up, have seen the cloud on which I was operating surprisingly reduced in magnitude.*"

Thoughts of a few Good Heads.

Emerson says : Nature makes fifty poor melons for one that is good, and shakes down a tree-full of wormy, unripe crabs, before you can find a dozen dessert apples ; and she scatters nations of naked Indians, and nations of clothed Christians, with two or three good heads among them. Nature works very hard, and only hits the white once in a million throws. In mankind, she is contented if she yields one master in a century. The more difficulty there is in creating good men, the more they are used when they come. I once counted in a little neighborhood, and found that every able-bodied man had, say from twelve to fifteen persons dependent on him for material aid—to whom he is to be for spoon and jug, for backer and sponsor, for nursery and hospital, and many functions beside ; nor does it seem to make much difference whether he is bachelor or patriarch ; if he do not violently decline the duties that fall to him, this amount of helpfulness will, in one way or another, be brought home to him. This is the tax which his abilities pay. The good men are employed for private centers of use, and for larger influence. All revelations, whether of mechanical, or intellectual, or moral science, are made not to communities, but to single persons. All the marked events of our day, all the cities, all the colonizations, may be traced back to their origin in a private brain. All the feats which make our civility, were the thoughts of a few good heads.

Curative Properties of Grapes.

Dr. Herpin, of Metz, has published a very interesting account of the curative effects of grapes in various disorders of the body. They act, first, by introducing large quantities of fluids into the system, which, passing through the blood, carry off, by perspiration and other excretions, the effete and injurious materials of the body; secondly, as a vegetable nutritive agent, through the albumeroid of nitrogenous and respiratory substance which the juice of the grape contains; thirdly, as a medicine, at the same time soothing, laxative, alterative, and defarative; fourthly, by the alkalies, which diminish the plasticity of the blood, and render all more fluid; fifthly, by the various mineral elements, such as sulphates, chlorides, phosphates, &c., which are an analogous and valuable substitute for many mineral waters. Employed rationally and methodically, aided by suitable diet and regimen, the grape produces most important changes in the system, in favoring organic transmutations, in contributing healthy materials to the repair and reconstruction of the various tissues, and in determining the removal of vitiated matters which have become useless and injurious to the system.

Dreams in Disease.

You must not deem all your "night thoughts" the whisperings of departed spirits. You are not in perfect health. In your present pathological condition, it is but reasonable to suppose that, at best, your "dreams are but mind-clouds—high and unshapen beauties—like mountains which contain much and rich matter." Bailey, in "Festus," said: "Dreams are rudiments of the great state to come. . . . . We dream of what is about to happen to us." But such dreams are the inspirations of the sleeper, when his bodily state is not much diseased, and when the slumber is nearly perfect and harmonious. Be not overmuch troubled.

Sleep, without Dreaming.

It is better (says a medical authority,) to go to sleep on the right side, for then the stomach is very much in the position of a bottle turned upside down, and the contents are aided in passing out by gravitation. If one goes to sleep on the left side, the operation of emptying the stomach of its contents is more like drawing water from a well. After going to sleep, let the body take its own position. If you sleep on your back, especially soon after a heavy meal, the weight of the digestive organs, and that of the food resting on the great vein of the body, near the backbone, compresses it, and arrests the flow of blood more or less. If the arrest is partial, the sleep is disturbed, and there are unpleasant dreams. If the meal has been recent or hearty, the arrest is more decided, and the various sensations, such as falling over a precipice, or the pursuit of a wild beast, or other impending danger, and the desperate effort to get rid of it, arouses us; that sends on the stagnating blood, and we awake in a fright, or trembling, or perspiration, or feeling of exhaustion, according to the degree of stagnation, and the length or strength of the effort made to escape the danger. Eating a large, or what is called "a hearty meal," before going to bed, should always be avoided: it is the frequent cause of nightmare, and sometimes the cause of sudden death

Patience, as a Medicine.

Patience and serenity of disposition should be cultivated *as a remedy*, to neutralize the irritability and fretfulness of the diseased organs. The electro-battery can do but little good. A faithful medium, when under the true Indian influence, can restore life to failing limbs. Sleeping with the head northward results in establishing the magnetic (or warm) forces in the vital system, and in directing the vital electricity (cold) upon the brain and cerebral nerves. Sleep, dreamlessness, and health, are the natural consequences. Always sleep with your mouth closed.

Summer Foods and Drinks.

The following sensible whisper is taken from Dr. Hall's *Journal:* Physiological researches establish the fact that acids promote the separation of the bile from the blood, which is then passed through the system, thus preventing fevers, the prevailing diseases of summer. All fevers are "bilious," that is, the bile is in the blood. Whatever is antagonistic of fever is cooling. It is a common saying that fruits are "cooling," and also berries of every description; it is because the acidity which they contain aids in separating the bile from the blood. Hence the great yearning for greens, and lettuce, and salads, in the early spring, these being eaten with vinegar; hence, also, the taste for something sour—for lemonades—on an attack of fever. But this being the case, it is easy to see that we nullify the good effects of fruits and berries in proportion as we eat them with sugar, or even sweet milk, or cream. If we eat them in their natural state, fresh, ripe, perfect, it is almost impossible to eat too many, or eat enough to hurt us; especially if we eat them alone, and not taking any liquid with them whatever. Also is buttermilk, or even common milk, promotive of health in summer time. Sweet milk tends to biliousness in sedentary people; sour milk is antagonistic. The Greeks and Turks are passionately fond of milk. The shepherds use rennet, and the milk dealers alum, to make it sour the sooner. Buttermilk acts like watermelons on the system.

Effect of Wearing Silk Dresses.

A lady correspondent propounds the following question: "*How does the wearing of silk dresses affect us ladies?*"

Answer.—The wearing of "silk dresses" exerts a variety of wonderful influences upon both body and soul. We have seen examples of intense chronic suffering occasioned by the habitual wearing of "silk dresses" *too tight* over the region of the diaphragm. Instances are on record, also, where *the length* of "silk dresses" has inveigled the wearer into divers and sundry difficulties. Deplorable cases are known where *the price* of "silk dresses" has disturbed the financial equilibrium of very respectable progenitors. That alarming and epidemical phenomenon of the age, known as *the trailing* of "silk dresses" over tobacco-stained pavements, is rapidly developing among sensible classes a psychological disease called "disgust." In young female minds we have observed, with some beautiful exceptions, that the wearing of very fine "silk dresses" produces an enlargement of certain cerebral organs—developing the symptoms of insulation, superiority to poor folks, pride, approbation, and temporary shallow-mindedness. The physiological effect of "silk dresses" is not much, however, unless the wearer is nervously-diseased and dreamful. Then the fabric is too electrical for health.

Corruption and Groans.

Mr. Emerson, in his late volume on the "Conduct of Life," says: "There is one topic peremptorily forbidden to all well-bred, to all rational mortals, namely, *their distempers.* If you have not slept, or if you have slept, or if you have headache, or sciatica, or leprosy, or thunder-stroke, I beseech you, by all angels, *to hold your peace, and not pollute the morning to which all the housemates bring serene and pleasant thoughts, by corruption and groans.* Come out of the azure. Love the day. Do not leave the sky out of your landscape. The oldest and the most deserving person should come very modestly into any newly awakened company, respecting the divine communications, out of which all must be presumed to have newly come. An old man, who added an elevating culture to a large experience of life, said to me: 'When you come into the room, I think I will study how to make humanity beautiful to you.'"

Cotton for Garments.

The best fabric for purposes of health is composed of flax and wool. Silk is a third-rate article for garments. It should be woven with linen or woolen, to supply the demands of physiological laws. Cotton is a neutral, or *passive* substance on the body; while linen is constantly electrical, and woolen powerfully magnetic. Silk, being a non-conductor of life and vitality, is inclined to exhaust the entire body through the nerve-forces. Linen and wool, woven into one fabric, are best for stockings, undershirts, and garments for the skin.

A Cure for Cold Feet.

If you have cold feet, immerse them morning and evening in cold water, rub with a rough towel, and run about your room till they burn. In one month you will be entirely relieved. This advice is excellent, and the remedy is good, "all other things being equal," but the truth is, that nothing external or internal can cure cold feet if the stomach does not promptly digest its daily food. A cheerful spirit, plenty of simple nourishment, whole shoes and cotton stockings, with appropriate outdoor exercise—these constitute the true medicine for cold feet. Zinc and copper plates, worn in the stockings at night, are very useful when your vitality is depressed.

Red and Black Pepper.

In the treatment of disease, black pepper is a powerful and useful irritant, and there are conditions of the stomach, bowels, and brain, in which pepper tea, or a few grains mixed with sugar, would act very beneficially. But it is not wise to use pepper every day. The effect is telegraphed to the membranes of the head, throat, and lungs. Pepper cannot be digested. It is foreign to the nature and composition of the blood, and is, therefore, propelled through the stomach and intestines by the coercive influence of the peristaltic motion. No doubt many persons have headache and bowel diseases from the constant use of such an irritant. Use no pepper, either black or red, except as a medicine.

The Wonders of Blood.

The liquid of the blood is colorless, and its red appearance is due to the presence of innumerable little bodies floating in it, which are so small that three millions of them are contained in a drop which may be suspended on the point of a needle. These corpuscles are sacs filled with a compound substance, and it has been ascertained what both the film of the sac and its contents are composed of. Each one of these little bodies has its own life. They are formed, and grow, and die; and it is calculated that nearly twenty millions perish at every pulsation of the heart.

Milk and Water as Beverages.

Children naturally love artificial beverages—especially the milk of goats and cows; and they love to imbibe the unfermented *juice* of fruit and berries, because such drinks are as natural for mankind as is pure water from the spring. In grape-growing countries men drink more wine than any other fluid; yet, beastly *intoxication* is a sad condition rarely seen in those regions. But of the future we prognosticate thus: *Wine, water,* and *milk* will eventually displace and banish alcoholic drinks, and all falsely-artistic table beverages will, in like manner, be swept from the earth; then, when sound health and common sense shall become the rule—and not, as now, the exception—the " wine-cup will be forever broken, and righteousness will everywhere prevail." "Fly swifter round, ye wheels of Time, and bring the welcome day."

Minerals in Vegetation.

M. Eugene Risler maintains that iron plays a principal part in the nutrition of plants; he shows that in the roots, seeds, and white portions, it exists as a protoxide, while in the green portions it is in the form of a peroxide. Expose vegetables to air and light, and the protoxide becomes a peroxide with a rapidity proportioned to the intensity of light. The chlorophyll is green because it combines the two oxides, blue and yellow; and they form a voltaic pair, which decomposes water and the carbonic acid held in solution, the carbon and hydrogen entering into the organism. Nocturnal nutrition is oxidation, diurnal nutrition is deoxidation, and the vegetable tissue is formed like the weaver's, night being the warp, day the woof, with the iron of the chlorophyll to serve as the shuttle.

Man in the Animal State.

"Do you think that man existed first in a germinal capacity in the first forms of animal life, and then grew up *through them*, developing more perfectly as he ascended?"

Our investigations bring us to this conclusion: that just as a Building exists, in a germinal state, in the mind of the architect first, then in all the materials accumulated, and lastly, in all the many and various forms which such materials are made to assume in the constructive process, until the idea is accomplished; so, in like manner, the design of a Man, male and female, was the original idea or conception in the spirit of Mother Nature and Father God; that this vast machinery of means (of minerals, vegetables, and animals,) are the appropriate materials of construction—through all which Man germinally lives, until fully organized as an entity.

Twenty-one Systems.

We have received many letters from students of physiology, who wish to read the old authors with light furnished by the Harmonial Philosophy. Much will depend on their own good sense and untiring industry, while studying. We would call the attention of all who love to study man's constitution—" fearfully and wonderfully made "—to the beautiful natural divisions of the textures of the corporeal organization. They may be divided into *three times seven systems*, thus :

1. Osseous.	1. Mucous.	1. Cellular.
2. Medullary.	2. Serous.	2. Nervous, *animal*.
3. Cartilaginous	3. Synovial.	3. Nervous, *organic*.
4. Fibrous.	4. Glandular.	4. Arterial.
5. Fibro-Cartilaginous.	5. Dermoid	5. Venous.
6. Muscular, *animal*.	6. Epidermoid.	6. Exhalant.
7. Muscular, *organic*.	7. Pilous.	7. Absorbent, with their glands.

A Medicine for every Home.

Not only should we cultivate such tempers as serve to render the intercourse of home amiable and affectionate, but we should strive to adorn it with those charms which good sense and refinement so easily impart to it. We say easily, for there are persons who think that a home cannot be beautiful without a considerable outlay of money. Such people are in error. It costs little to have a neat flower-garden, and to surround your dwelling with those simple beauties which delight the eye far more than expensive objects. Nature delights in beauty. She loves to brighten the landscape and make it agreeable to the eye. She hangs ivy around the ruin, and over the stump of a withered tree she twines the graceful vine. A thousand arts she practices to animate the sense and please the mind. Follow her example, and do for yourself what she is always laboring to do for you.—*Colton.*

Wretchedness at Home.

It is not a medical whisper, but a short sermon on Love, and Trust, and Faith, which many require for bodily recuperation. Perhaps you have lost the bright fresh feelings of the soul. If so, you fancy your body diseased. Stand straight up before yourself, then, and let the native power of your soul shine and work into your daily life.

> "Oh, it is great to feel we care for nothing—
> That hope, nor love, nor fear, nor aught of earth
> Can check the royal ravishment of life;
> But like a streamer strown upon the wind,
> We fling our souls to Fate and to the Future.
> ———On ! said God unto the soul
> As to the earth, forever. On it goes,
> *A rejoicing native of the infinite*—
> As a bird of air—an orb of heaven."

We administer the foregoing as the *remedy* best suited to heal many infirmities; the balm for many a broken head and heart; the best medical whisper in our pharmacy for hundreds

The Pleasures of Home.

Question.—A correspondent asks : " Suppose all marriageable persons should, from this day, marry in accordance with what you term 'temperamental adaptation,' do you think the *ordinary* broils and vexations of human homes would cease altogether ?"

Answer.—If the writer had made a more comprehensive supposition (including all the married throughout the world,) we should reply affirmatively; excepting, of course, all such ordinary " broils " as those which are indispensable to meet the demands of honest hunger.

Let the already truly married *practice* the principles of Harmonial love and wisdom toward each other, and let those about to embark upon the conjugal existence, regulate their attachments and lives by spiritual delicacy and private truthfulness. We can guarantee that such a home will be a natural Sanctuary of heavenly blessedness. "The family circle" would shine like a ring of diamonds. Each throbbing heart would be a well-spring of love, tenderness, grace, and gladness. Good angels would go in and out of such a sunny home, just as the healthy children thereof would glide to and fro, on the swift feet of unrestrained enjoyment. A divine joy is certain to pavilion such a happy home, and one tender bond is sure to embrace all hearts; for it would be the royal house of the indwelling God, and the very "gate of heaven."

Sympathy with an Amputated Limb.

It is a truth that man's soul does not leave an amputated limb and take up its residence in the interiors of the living body until after the lapse of many days; even though, to all external sense and seeming, the violently removed and buried member would indicate no possession of life. The brain maintains a nervo-vital record and governmental sympathy with each part of the lower organism, and such sympathy cannot be withdrawn violently nor hastily. After death, as we have frequently observed, the vitality treasured up out of a limb severed from the body, elaborates the spiritual limb in exact harmony with the form and proportions of what Nature had first decreed. So that, in the immortal state, each maimed and deformed individual appears, not as he externally looked at the moment of death, but, instead, in such form and embodiments as he would have possessed if Nature's designs had not been arrested in their development in this world.

An authentic instance of cerebro-vital sympathy with an amputated limb, will illustrate the principle. Recently, at Tower's Mill, in Lanesborough, Mass., a young man named Jerry Swan was caught by the arm in some machinery, and the limb was so badly broken and mangled that immediate amputation was necessary. This was successfully performed, but, according to the Pittsfield *Eagle*, Mr. Swan's connection with the dissevered limb did not cease with the operation. The *Eagle* says: "On recovering from the stupor (produced by the use of chloroform,) Mr. Swan still complained sorely of an aching hand. Late in the evening his distress became very great, and he insisted that the hand was cramped by being doubled up. The limb had been placed in a small box and buried. His attendants dug it up and straightened the hand, and he was soon easier. This morning the limb was again buried. But he soon complained of a sensation of cold and great pain in it. It was accordingly taken up again, wrapped up, and deposited in a tomb, since which he is again relieved."

The Influence of Indian Spirits.

Indian spirits are robust, healthy, and sympathetic; but they seldom confer *wisdom* upon their mediums. In the Spirit Land they are exceedingly officious and useful in many ways; particularly in receiving and taking sympathetic charge of the spirits of persons who have just died in hospitals, by accident, or on the field of battle. They exhibit the finest shades of sympathy and brotherly love, but are rarely wise and prudent in the employment of their powers. For this reason, principally, the gregarious tribes of the Spirit Land are subdivided into classes, as in a school; and thousands of illustrious wise men, once so called on earth among men, delight in appointing themselves to the office of *monitors* and *teachers* among the classified Red men who are so grouped in the celestial spheres.

State of the First Man.

QUESTION.—"Was the first man created in a state of infancy, or did he emanate from the Divine Hand in a high state of intellectual development?"

ANSWER.—Whatever is right and authoritative derives its sanction and power, not from popular opinion or statute law, but from the true order and harmoniousness of the universe. Divine revelations, so called, are nothing, unless they coincide with the teachings of Nature, whence such revelations derive whatever of truth and authority they may inherently possess. The teachings of Nature are explicit to this point: That the "first man" was born just as the first child in any family is born (*while exceedingly young,*) and that he was not "created" in a perfect state of intellectual development, but commenced the journey of life crowned with every glorious endowment, yet clothed with ignorance or inexperience.

The Caging of Birds.

We contemplate the operations of our loving Mother, and solicit instruction from her million voices. From every mouth we hear the electric word—"Liberty!" A bird is a beauteous bit of ascending Nature on wings of flight. Man is another form of the same Nature, walking upright, the Lord of all beneath his exalted mind and commanding position. He is good and glorious in his government when the spirit of freedom prevails in the least of things about him; but what shall we say of his lordship, of his influence and administration, when he cuts the wings of Liberty, cages the birds of paradise, and enforces obedience by his will with a rod of iron! Judge of yourself what is right.

Spiritualistic Superstition.

QUESTION.—"You have, on at least one occasion, hit Spiritualists pretty severe raps for being 'superstitious.' By unbelievers, *all* Spiritualists are supposed to be superstitious. Will you please define what you mean?"

ANSWER.—We denominate all persons superstitious, who, with excessive credulity, and no absolute evidence, attribute unusual physical sensations to the work of spirits; also, they are superstitious who imagine that their own mishaps and discords were developed by the special intervention of the invisible. It is superstition to believe that a medium is influenced by any disembodied intelligence to do or say anything earthly and sensual. We hold every one individually accountable for all unworthy speeches and physical indulgences. Let all Spiritualists believe that "every good and perfect gift cometh from above," and reversely, that every discordant and miserable influence is generated beneath, in the regions of the flesh, and we will assure them that the "mediums for evil spirits" will number far less than at present.

True and False Hospitality.

We think that if friends and strangers would deal with each other candidly and gently, without dissimulation and hypocritical etiquette, they would never be driven from each other by distrust and enmity. It is the most malignant form of hypocrisy to invite persons to call upon you, while, secretly, you wish they would not; and it is, perhaps, not less injurious to extend the hospitalities of your house to individuals, while, in truth, you dislike them, and want them removed from your vicinity. But there is a higher development for you and all, namely, to feel the spirit of universal adoption, whereby, at all times, you can impart *hospitality* and *kindness* alike to friend and stranger, without violating the principles of self-justice and truth. It is, however, far better to be honestly and undisguisedly exclusive, than to be hypocritically hospitable for the purpose of acquiring a reputation for benevolence and philanthropy.

Would You Stop the Flowing River?

Many correspondents will find our answer in the following lines:

> "He who checks a child with terror,
> Stops its play and stills its song,
> Not alone commits an error,
> But a great and moral wrong.
>
> "Give it play, and never fear it,
> Active life is no defect;
> Never, never break its spirit—
> Curb it only to direct.
>
> "Would you stop the flowing river,
> Thinking it would cease to flow?
> Onward it must flow forever;
> Better teach it where to go."

The Duty of the Skin.

Some curious facts were presented in a recent lecture of Dr. Thudicum on the Turkish Bath. The human body can bear 300° of heat. The perspiration from a clean skin has an agreeable odor or none at all, while a disagreeable one is the product of an ammoniacal salt, formed of urea and volatile acid. The ventilation of the bulk of tissue—cellular and muscular—is the peculiar duty of the skin.

Traveling as a Medicine.

The principles inculcated in our philosophy of Disease, are within the intellectual grasp of every reader; and we think that every patient is physically qualified to apply our rules of treatment. The scientific *name* of your disease is of no consequence, so far as the application of this philosophy is concerned. The state or symptoms are the all-essential questions. Life is worth nothing without *Health* as its crowning glory, and even this glory is worthless without those spiritual feelings which exalt and beautify existence. Rely on your Self-Healing energies! Traveling is vastly better than medicine for every one who cannot break up a tedious *routine* of habits and feelings at home.

Vegetarianism among Animals.

There are many animals that practice nothing but pure Vegetarianism, and yet, as characters, they do not improve. Linnæus states the cow to eat 276 plants, and to refuse 218; the goat eats 449 and declines 126; the sheep takes 387, and rejects 141; the horse likes 262, and avoids 212; but the hog, more nice in its provision than any of the former, eats but 72 plants, and rejects 171.

A Word to Magnetizers.

If you want wisely to affect your patients, do not talk to them while bestowing the benefits of your manipulations. You cannot ask questions, and keep an intelligent conversation with your patients, while exerting yourself to overcome their diseases. You need all the *breath* you can inspire, while your patient needs a perfect tranquillity of the mental organs. Practice with the strictest composure, with ceaseless assiduity, with loving kindness, with the sweetest devotion to the cause of Health and Happiness.

Infusorial Reproduction.

The polypi are reproduced just as the blood globules and most cellules in animal bodies are formed—namely, by buds or eggs cropping out from the vital parts, attaining to perfection, and then dropping off somewhat as apples fall from their producing boughs. Among the infusoria there is no direct organic contact for purposes of reproduction; for, like most medusa and worms, each individual is hermaphroditic, having the reproductive organs of both sexes.

Value of Sunlight in Houses.

The following fact, says a good authority, has been established by careful observation ; that where sunlight penetrates all the rooms of a dwelling, the inmates are less liable to sickness, than in a house where the apartments lose its health-invigorating influences. Basement rooms are the nurseries of indisposition. It is a gross mistake to compel human beings to reside partially under ground. There is a defective condition of the air in such rooms, connected with dampness, besides the decomposing paint on the walls, and the escape of noxious gases from pipes and drains. It is strange that builders persist in doing violence to humanity, by still erecting houses with basements.

Enlarge your Thoughts and Perceptions.

Perhaps your experience is replete with items of imperishable wealth. Never expect inspirations of truth, however, until you ascend to the altitude of Principles. The eye that permanently contracts its scope to the points of diamonds, however valuable in themselves, is not capable of surveying the chemistry of elements, nor is such a mind capable of feasting upon that happiness which results from the grander and freer employment of the intellectual powers.

We hope and expect to impress hundreds of thousands with the belief that *there is no antagonism between enlightened Reason and Nature's highest, most central Truth, which is Father God.* If mankind will reverently learn the lessons of Nature, and become intelligently receptive, like the unconscious flower which unrolls its petals to receive the vivifying heat and light of Heaven, they may very rapidly bud and blossom into happy families and progressive Brotherhoods.

What and Where is Heaven?

Heaven, as we use the term, signifies the soul's Harmony with Deity; so that, in very truth, such a person can say: "*I and my Father are one.*" This, and this only, is HEAVEN. Country, climate, time, space, infinity, eternity—these have nothing to do with the soul's true Heaven. Wherever the human spirit beats thankfully and melodiously with the pulsations of the Soul of Nature, *there*, in that spirit's silent deep, is the kingdom of peace and righteousness.

Yet, in the outward immensities of the universe, where the stars are suns of immeasurable magnitude, and where myriad groups of them form circles inconceivable, there are substantial worlds, or *locations*, which we term "Spiritual Spheres."

How to Quiet Children.

Let them have a goodly supply of bodily activity during the day-time. Give them nothing *sweet*, except your smiles, after the second meal. Avoid domestic contradictions in their presence, and never give way to anger in their absence. Never give a child either tea or coffee. Many little ones crave and ask for stimulants, because of hereditary bias and the force of daily example. If your child is very irritable, give its body a good rub-down, as you would dress the limbs of a young colt, just before delivering it over to the safe keeping of the guardian Angels of the night.

Don't Stand on the Track.

The train, says a railroad gazette, may steal suddenly upon you, and then a little trepidation, a slight misstep, a slip of the foot, and we shudder to think of your crushed and bleeding body. So, in the journey of life, perils are around you on every hand. But don't plant yourself in their path and defy them;

don't stand in their track and disregard them. Perhaps you now and then take a little intoxicating drink. My friend, if so, you are standing "on the track," while the car of retribution comes thundering on—moving in a right line—approaching with steady and rapid wheels. Will it not bear down and crush you? Perhaps you spend an occasional evening with a party of friends, amusing yourself with cards or dice, staking small sums to make the game interesting. My friend, you are standing "on the track." Thousands have stood there and perished. Don't wait to hear the panting of the iron steed, and the rattling of rushing wheels, but fly from the track. At a safe distance, stand and view the wrecks which yon ponderous train will spread before you. Look well to the ground on which you plant your feet, and forget not, for yet these many days, our parting words : " Don't stand on the track."

Pure Alcohol as a Medicine.

A correspondent writes for information respecting the effect of pure alcohol on a scrofulous system—especially as a remedy in cerebral embarrassments—occasioned by a humor partly inherited, and aggravated by a diseased condition of the liver.

Answer.—The human system cannot appropriate the liquid known as alcohol. Years after it is imbibed, you will find it in the serum and other fluids of the brain, and in precisely the same state as when it was originally consigned to the stomach. If the blood would accept of alcohol as a component part, or if the solid and tissual structures of the body would temporarily convert the liquid into their uses, then the voice of wisdom would proclaim the utility of an alcoholic remedy for disease. But all this is impossible. The whole body, under the government of the brain, says to alcohol : "Monster! stay without the temple." And yet, we know that there are circumstances and conditions of the brain and nerve-system, when alcohol, in a modified form, is one of the best medicines within the reach of man. Snake poisons, and some forms of insanity, will yield to its counter-irritations. But let no man expect a better brain, a finer digestion, or a more active liver, from the use of forbidden fluids.

The Disgusting Habit of Chewing.

Go where you will, says Dr. Dixon, in his *Scalpel*, on board of any steamboat or rail-car, and you witness the disgusting habit of chewing every variety of unhealthy mess, nuts and candies, that may be offered by traveling hucksters, and immediately it is transferred to some human mouth. It would almost seem that everybody in America is half starved; the chewing, munching, spitting, and tooth-picking, has become disgusting to every decent foreigner. Nature requires food periodically; there is no mistaking the call; when hungry, the stomach and salivary glands are prepared for it. As a general rule, instinct indicates what is wanted, and when it is wanted; and whoever habitually eats whenever food of any description is placed before him, will never reach middle life in health. Even animals know better, for when their hunger is appeased they will stop. What a poor slave is he who has learned to chew tobacco! He must always have a big box in his pocket, a big quid in his mouth, and a big dish for the juice. Poor fellow! if he had no company in this filthy habit, everybody would believe him crazy. You say you can leave it whenever you choose. Then clean out your pockets, your mouth; get your linen purified and bleached; burn cotton till the disgusting odor is removed, and quit.

The present annual production of tobacco has been estimated by an English writer at 4,000,000,000 pounds! This is smoked, chewed, and snuffed. Suppose it all made into segars, 100 to the pound, it would produce 400,000,000,000. Four hundred billions of segars! Allowing this tobacco, unmanufactured, to cost on the average 10 cents a pound, and we have $400,000,000 expended every year in producing a noxious, deleterious weed. At least one and a half times as much more is required to manufacture it into a marketable form, and to dispose of it to the consumer.

The Madness of Intemperance.

Judge Johnson, in sentencing a culprit to death, said : "Nor shall the *place* be forgotten in which occurred this shedding of blood. It was in one of the thousand ante-chambers of hell, which mar, like plague-spots, the fair face of our State. You need not be told that I mean a tippling-shop, the meeting-place of Satan's minions, and the foul cesspool which, by spontaneous generation, breeds and nurtures all that is loathsome and disgusting in profanity, and babbling, and vulgarity, and Sabbath-breaking. I would not be the owner of a groggery for the price of this globe converted into precious ore. For the pitiful sum of a dime, he furnished the poison which made the deceased *a fool*, and this trembling culprit *a demon*. How paltry this price of two human lives ! This traffic is tolerated by law, and, therefore, the vender has committed an offense not cognizable by earthly tribunals ; but, in the sight of Him who is unerring wisdom, he who deliberately furnishes the intoxicating draught which inflames men to anger, and violence, and bloodshed, *is particeps criminis* in the moral turpitude of the deed. Is it not high time that these sinks of vice and crime should be held rigidly accountable to the laws of the land, and placed under the ban of an enlightened and virtuous public opinion ?"

The Evils of Opium-eating.

The evils of the too prevalent habit of opium-eating are well illustrated by a case recently reported. Dr. Kozieradzki, in a foreign medical journal, says that he was called to see a lady aged thirty, who had suffered since her thirteenth year from a painful affection of the heart, and had obtained ease by taking opium, increasing the dose from one-eighth grain to thirty grains, so that, at last, two hundred grains were taken daily. After a while, epileptic fits came on. She had a prematurely old appearance, was deaf of one ear, and had lost almost all her teeth. Being one day unable to procure her accustomed dose of morphine, and having to pass twenty-four hours amid pains at the heart and repeated fainting, she became so much prostrated that, when next day her accustomed dangerous anodyne was brought her, she was seized with a convulsive attack, and died.

The Opium-eater's Insanity.

Nothing but a potent determination to abandon the use of the opium will effectually cure it, and compensate for its customary effects. If there be a venous enlargement, extending upwardly to the brain, the fact is against an independence of the drug. You may pacify sometimes, by applying fomentations of bitter herbs to the stomach; and, internally, a teaspoon twice full of the tincture of hops in a little common black tea, without milk or sugar.

Substitute for Strong Drink.

A correspondent asks for information respecting the nature of intoxication, and whether there is any remedy or substitute for it. The case is that of a respectable and intelligent man, who, unfortunately, is a great lover of rum, and often gets beastly drunk. His passion for strong drink is more powerful than his will, and his slavery is, therefore, absolute.

REMEDY.—All attacks of disease and discord are periodical. They are developed and governed by the retirement of one set of forces and the ascendency of another; so that all remedial enterprises must be instituted with strict reference to *the* week (and particularly to *the* day of the week,) when the patient usually yields to "the passion for strong drink." Passion is not as a sleeping lion aroused from his lair by the presence of temptation, but it is the reaction of a class of discordant faculties—and this explains why some persons experience and demonstrate *anger* without any sufficient apparent cause, and explains also why many very noble characters yield to the ruthless government of some silly weakness far beneath their moral and intellectual *status*. Of course, no patient can be cured of his "passion for strong drink," unless his whole soul be fully inspired with the sublime wish to be *a whole Man!* Take this as the bottom-work of the foundation, then watch the period of habitual intemperance, and forestall it by shielding the appetite in every conceivable way—by withholding the customary table temptations. This system should commence at least three days before the "passion" is stealthily developed to its overmastering hight. When the patient (commonly termed "the drunkard,") is impassioned and mad for his intoxicating beverage, administer the following:

Extract Hasheesh (*Indian Cannabis,*) 15 grs.

Sweet Flag-root (*Acorus Calamus,*) pulverized perfectly, 1 scruple.

Mix well, and divide into fifty powders. Give one every two, four, or six hours, according to the intensity and incorrigibleness of the victim. In some cases a much *finer* preparation would be required. Once pass the season of temptation, escape the period of passion, and the battle is more than half-won. But there is another remedy for this malady.

An English Cure for Drunkenness.

There is a famous prescription in use in England (says the *Springfield Republican*,) for the cure of drunkenness, by which thousands are said to have been assisted in recovering themselves. The recipe came into notoriety through the efforts of John Vine Hall, father of the Rev. Newman Hall, and Captain Vine Hall, commander of the Great Eastern steamship. He had fallen into such habitual drunkenness, that his most earnest efforts to reclaim himself proved unavailing. At length he sought the advice of an eminent physician, who gave him a prescription which he followed faithfully for seven months, and at the end of that time had lost all desire for liquors, although he had been for many years led captive by a most debasing appetite. The recipe, which he afterward published, and by which so many other drunkards have been assisted to reform, is as follows: "Sulphate of iron, five grains; magnesia, ten grains; peppermint-water, eleven drachms; spirit of nutmeg, one drachm; twice a day." This preparation acts as a tonic and stimulant, and so partially supplies the place of the accustomed liquor, and prevents that absolute physical and moral prostration that follow a sudden breaking off from the use of stimulating drinks. In cases where the appetite for liquors is not too strong, the medicine supplies the place of the accustomed drams entirely, but Mr. Hall continued the use of liquors at first with the medicine, diminishing the amount gradually until he was able to throw away his bottle and glass altogether, after which he continued to take the medicine a month or two, till he felt that he was wholly restored to self-control, and could rejoice in a sound mind in a sound body.

Physical Evils are Transient.

An uncle of mine (says a correspondent,) aged eighty-five residing here, after having been a habitual tobacco-chewer for *more than sixty years*, about two years ago lost all appetite for it, and has had *no desire* to taste it since! There was no change in his habits, and he was always more healthy than most men. Some Spiritualists say that the *love* of the tobacco, alcohol, &c., adheres to man in the spirit-life. But may it not be, as in this instance, that, by some peculiar *chemical* or physiological change in the organism of the spirit, the appetite may leave them immediately on their introduction into the other life?

Natural and Artistic Beverages.

Man is an artist, a natural inventor, and his intellectual endowments, urged by the inward force of his diversal wants, among other necessities and inventions, impel him to compound his foods and to artificialize his drinks. It is, therefore, strictly natural for man, while passing through the *transitions* from a crude Savagism to a spiritualized Civilization, to drink and eat many hurtful (because artificial) inventions of his own unfolding genius in the direction of art and discovery. But the primordial and instinctively simple beverage is *pure spring water*. It is the most unartistic, the cheapest, most poetic, altogether beautiful, and the healthiest.

Man's Voluntary Powers.

Our philosophy teaches us that man, beginning his earth-life as an automatic being, will, when ultimately unfolded in all the hidden centers of spirit, become wholly *voluntary* · that then, solely through the energetic fiat of his untrammeled · will, he may direct or withdraw vital forces to and from any part of his organization. In this way, we hold that man will " heal himself" when afflicted with disease, and shield his body from the ruthless assaults of epidemics and contagious disturbances. In short, we believe that man is organized to triumph over all his enemies.

At this proposition, the doctors level their shafts of ridicule. "I scout the doctrine," says one, "because the mind's voluntary powers are fixed within certain well-defined and unalterable limits," &c.

We reply, that no intelligent physiologist can venture to fix the domain of mind without some knowledge of psychology—a department of science as yet scarcely known to our best medical scholars. Man is unspeakably superior to fish, bird, or beast; and is endowed with powers greater than all the millions below him. The voluntary powers of the *inferiors'* brains are but partially understood.

The Alpine hunter will tell you that the Chamois, a beautiful creature among the everlasting mountains, is capable of running at the rate of one hundred and twenty-five miles per hour.

The Ostrich will run swifter than the fleetest horse.

Certain fish seem to fly rather than swim.

Some birds can dart against, and through, the adverse aerial currents, and will make nearly two miles per minute, showing that they might, if kept in one course, encompass the earth in less time than is required for a fast steamer to cross the Atlantic.

The crushing might of a cannon-ball is less than the awful force with which a Whale strikes the ocean's bosom. The powerful drumming of the Gorilla upon its own breast can be heard at the distance of half a mile. Some muscular animals, like the Buffalo, can dive from ten to twenty feet under water.

And all these exhibitions of velocity and strength, take their rise from *the voluntary centers of the animal's brain.* Is not man destined to be, and to do, more than any creature which is inferior to his exalted make and station? We believe that he is—do you still doubt, reader?

The Mind in Sleep.

QUESTION.—"Do you think that the mind, in sleep, is independent of the body? If so, does it wander in the Spirit Land? If such be the case, why is it that spirits are supposed to be taken by surprise when they die here, and waking, find themselves in a strange place?"

EXPLANATION.—Our answer in this place must be brief, more particularly because the subject has been considered in the third volume of the Harmonia.

Notwithstanding all our investigations in the realms of mind, we have no knowledge of an instance where spirit escaped the body, except by means of thought, idea, or consciousness, until the moment of final dissolution. But through the medium of idea, feeling, or *clairvoyance*, the apparent escape and the *seeming* independence of the spirit, are enough complete to impress both subject and witness with a conviction of absolute certainty. We held this belief for nearly three years.

The philosophy of mind is the same as the philosophy of a river of pure eternal water. Man's spirit is a substance, composed of all the divine essences and principles. Its consciousness is the effect of motion, just as the waves of water are the results of some disturbing force. The mind may be quite *still* during sleep; therefore, during sleep, it may be wholly *unconscious*. If the substance of the mind is in motion during sleep, it will then, of necessity, be both conscious and dreamful. It may, however, be exalted into clairvoyance, so that, through *sight* (and even through all the spiritual senses,) the mind may extend its personality, or *consciousness*, into new scenes and associations.

Electricity and Phosphorus in Animals.

An investigator of the faculties and habits of certain fish and insects, asks to be aided " by a few impressions on two points : First, whether the lightning-bug is endowed with a power of collecting and discharging electricity the same as the torpedo or electric eel, and second, whether this power is functionary or under the control and direction of the fish's and insects will ?"

Answer.—These questions reach down and jut out into the realm of particulars. Our impressions may be too *general* for your purpose, but they are at your service. The lightning-bug, so called, is capable of generating and emitting light from the presence of that mysterious substance which was first discovered by the Alchemists, but now everywhere known as *Phosphorus*, an essential element of all animal organization. The generation and employment of this phosphorescent light seems to be subject to the insect's will. There are many impressive instances on record where this fire-fly has destroyed poisonous spiders and consumed their webs, when the necessity of self-defense urged the little creature to persistent action.

But nothing of this is true when we come to examine the means of self-defense peculiar to the torpedo, ray-fish, electric eel, etc. With these beings of the ocean, the truth seems to be, that, by means of their digestive membranes, they can generate and accumulate the electric fluid in large quantities. This they do with wonderful rapidity when excited by either fright or anger. The under-skin becomes distended and puffed like that which holds the quills of the "fretful porcupine." Thus, the electric eel is charged like the Leyden Jar, and can, unlike the jar, emit sparks without coming in contact with any other body. It is now prepared to wage an aggressive war, or to do battle in obedience to the instincts of self-preservation. Fishes of this class can, at will, emit sparks in all directions, can give off brilliant emanations, which often prove fatal to the recipient, and can direct the lightning of their privately-forged thunderbolts, which seldom fail of hitting the mark. The electricity employed is identical with that formed in the vital centers of all animal organisms.

Causes of Hemorrhoidal Infirmities.

More than two-thirds of the sick who write to us for prescriptions are afflicted with *hemorrhoidal* troubles, commonly called "piles," of which there are many forms and painful symptoms.

THE CAUSE: Pathological physiologists have for a long period said that piles are produced by a dilatation of the bloodvessels in the walls of the rectum; that the cellular coatings of the lower orifice, by becoming enlarged and flabby, protrude and discharge mucous or blood; and with this explanation, they have classified piles as (1) *blind*, (2) *mucous*, (3) *bleeding*, (4) *excrescential*.

The first form is known as a thickening or swelling of the membranes and vessels within the rectum; the second form is a discharge of mucous from what appear to be ulcers, but which is usually nothing more than a pus exuded from excoriated surfaces within the anus; the third form, attended with pain and uneasiness, is characterized by a discharge of blood during evacuation; the fourth form, and by far the most difficult to treat, is known by the existence of fleshy tumors in the upper walls of the rectum, flat or fig-shaped excrescences, which are commonly removed by surgical operation. But we perceive a better explanation.

The *causes* of piles begin at the brain—in the center of all energy. They signify an unbalanced condition of the nervous system. The registration of this nervo-disturbance is made upon the negative side of the circulating system, namely, upon the veins, and the local development of the condition is invariably wherever the body is the weakest. The disturbance in one person may be called "liver disease;" in another, "costiveness;" in a third instance, "apoplexy;" or "piles," if the rectum and lower bowels are the weakest or most taxed part of the venous system. That piles originate from mental rather than from physical disturbances, may be easily demonstrated. Piles may result from *anxiety, or sorrow, or suffering*, or from long-continued *excitement* of the feelings in any direction. Straining and anxiety during pregnancy and parturition have frequently brought them on; or occupying the mind with reading papers or books while engaged in the function of evacuation. This habit is as dangerous to the lower intestines and rectum as it is disgusting and offensive to every fine sensibility. It is a very common cause of piles. Any *mental* occupation foreign to the proper and prompt performance of the function, is positively certain to stamp the impress of disease upon the weakest part; and, inasmuch as, while engaged in this particular function, the vessels and fibers of the rectum are distended and principally taxed, so is inattention at the time most likely to produce one or more of the above-mentioned forms of hemorrhoidal disturbance.

It should be remembered that the veins are negative to the arteries, and that *mental* disturbances are more likely to telegraph themselves upon the most negative parts of the venous system; which explains why little children, and even animals, are sometimes victims of piles and diseases of the anus. And it is worthy of very particular remark, that piles are *periodical* in their appearance, painfulness, and disappearance. If, for example, you suffered most from them in April, last year, you will be very likely to experience a return of similar troubles about the same time this spring. Or, if you suffer most from piles in the after part of the day, the symptoms will revive during the same hours to-morrow; thus giving your mind the impression that *periodicity* is a law as much in disease as in health.

REMEDY.—The treatment recommended for this disease, is different in different schools of medicine. The remedies are almost innumerable, and yet, the disease prevails throughout civilization. Thousands have the *piles* without whispering the fact to their nearest friend. People seem to regard with indelicate suspicion, if not with mortification, certain parts and functions of their organization. Excessive modesty is not the sign of intelligence and refinement, but it is, rather, a symptom of conscious embarrassment—a momentary loss of self-possession—while in the presence of those who are supposed to think unworthily of the subordinate functions of physical organs. Intelligence and refinement, instead of masking and falsifying themselves in furtive glances and prudish expressions, will throw a white halo of significance and respect around the least of things.

Physicians of every school, when they ascertain that their patient is afflicted with piles, will prescribe, in addition to physical treatment, *mental* quietude and freedom from muscular exertion. The disciples of Hahnemann have found sovereign remedies for piles in their minute pellets and powders, because they very judiciously started with the theory that the disease is a *spiritual* disturbance; and hence, logically, that prolapsus ani and hemorrhoids will yield to the *spiritual* parts of Nux vom. and Sulphur, Belladonna, Arsenicum, Carbo-veget., Mercur. sol., Ignatia, Cocculus, or to the active principle of some other remedy or combination in the pharmacopœia of infinitesimalism. All the success of homeopathic physicians, according to our perception, is attributable to the practical application of

21*

a correct theory in their possession, that *disease is spiritual in its origin*, and that symptoms, consequently, are the voices of the internal disturbances, which may be hushed and forever silenced by the prompt administration of whatever they (the voices,) infallibly call for—this, more than the "like-cures-like" principle, is the central secret of the triumphant progression of Homeopathy. But more of this hereafter, when we come to write a few articles on the "isms" in the world of Medicine.

In order to treat piles, or prolapsus ani, successfully, it is necessary to obey all the psychological by-laws, and comply with all the mechanical conditions which the disease suggests to the understanding. First, you must cease straining to discharge the excrementitious contents of the intestines; second, you must conform to rules heretofore given, by which "costiveness" is effectually cured without medicines; third, you must not occupy your thoughts and dissipate your Will-power while performing your bodily functions; fourth, you must not procrastinate the hour of obedience to this demand of your being, but the utmost system and regularity are required, and if your daily labor calls you early, so that you cannot attend to this function, then adopt the hour before bed-time as the most appropriate, and insist upon teaching your intestines to respond promptly at this period; fifth, if the vessels or lips of the anus protrude after the operation, lean forward and push them fully *upward* within the orifice, to their proper position. Never neglect this mechanical adjustment of the fallen and distended vessels, for there are reasons why the contracting muscles cannot always accomplish this important result. If the vessels are allowed to remain without the rectum for a few days, they become *strangulated* and hardened—perhaps ulcerous and exceedingly sore—

so that a surgical operation is sometimes the only way of escape from results more serious.

Ointments and salves are *not* curative. They seldom do anything more than soothe the external and exposed parts, while they almost certainly inflame the cellular membranes of the rectum. The drug stores are full of quack medicines for piles, hernia, &c., not one of which is likely to accomplish more than temporary relief. One of the best palliative treatments is that of our hydropathic establishments—correct diet, frequent bathing, and cold water syringes. But the cure, whenever accomplished, occurs by and through the restoration of the primeval equilibrium between the *nervous and the venous systems*, or, more properly speaking, by establishing a balance of the spiritual forces (dynamics,) in the circulatory organism, which includes every possible ramification of the veins and arteries, beginning in the brain, pouring through the *moderator*, called the "heart," and diffusing its life-principles to the minutest bounds of the structural economy.

Recently our attention was arrested by the assertion of a scientific gentleman in this city—one of our first inventors—who declared that, without being able to give a philosophical reason, he had been *cured of piles by simply carrying a horse-chestnut in his pantaloons' pocket*. The discovery of this peculiar power in the horse-chestnut was accidental on the part of another gentleman, who had been previously cured. It was found that, so long as he carried the chestnut about his person, the piles, although exceedingly annoying and painful before, were kept in subjection, and at last entirely overcome.

We have examined the emanations of the horse-chestnut since the above fact came to our knowledge, and do not hesitate to prescribe the experiment to our patients. Indeed, we do not recommend it as an "experiment," but as a *remedy;* inasmuch as the active principle of the nut is friendly to the removal of hemorrhoidal swellings; but it is not a remedy under violations of by-laws and conditions already specified.

The subtile penetrations of certain metals and plants are exceedingly curious and magnetical. Much superstition and folly have been developed by too much reliance upon the virtues of various herbs and mineral bodies. And yet, there is a scientific truth at the bottom of all mythology, alchemy, and astrology. The ancients were peculiarly, but often erroneously, impressed with the spiritual properties of salts and herbal preparations. Alchemists, for example, found much spiritual (*i. e.*, dynamic and psychologic force or) energy in what they termed the "Volatile Salt of Vipers." One writer says: "To judge well of the effects which this Volatile Salt can produce in our bodies, we must know its manner of operation, which is to open, to comminute, to pierce, to attenuate, and to drive to the extreme parts of the body, and through the pores of the skin, all the impurities and all the strange bodies that can get out by those ways. Further, it is an enemy to all corruption—very friendly and very agreeable to our nature, which it assists and fortifies, enabling it to expel, not only by the pores of the skin, but by siege, and by all the emunctories of the body, the superfluous humors which molest it; whence it comes to pass, that it produces admirable effects upon a thousand occasions, curing a great number of sicknesses, or, at least, giving great relief therein, even in those that are most refractory and most difficult to cure—such as apoplexies, lethargies, convulsions, agues, and many other maladies, believed to have their source in the *brain*."

With this singular quotation from one of the ancient alchemists, who was evidently impressed with the *aural* powers and odyllic energies of mineral bodies, we will conclude—hoping to hear from correspondents after carrying the horse-chestnut for a few weeks. It seems to us that the constant presence of the chestnut for the cure of particular diseases, will arouse and concentrate your pneumogastrical powers.

Piles Cured by Horse-chestnuts.

FREMONT, Ind., June, 1861.

To A. J. DAVIS: Recently, in conversation with a gentleman (a mason by trade,) he stated that, some years since, he was attacked with the piles, which readily disappeared, in consequence (as he verily believes,) of carrying a *horse-chestnut in his pocket*, and continued well for about five years; during which time he carried it (the chestnut,) in his pocket; becoming worn out, he threw it away. The disease returned in a few weeks, but was again promptly removed by resorting to the same remedy, which he still uses.

He also informed me that an acquaintance of his (also a mason,) related to him a similar experience respecting the sanative influence of this apparently very simple remedy.

Truly yours,   S. W. CORBIN, M. D.

The Cause and Cure of Hydrophobia.

"MY DEAR SIR: Please inform me, as soon as you possibly can, what is the best medicine for preventing or curing Hydrophobia. Mad dogs have been lately very numerous through this neighborhood. Do, for humanity's sake, inform your readers what is the most efficacious treatment of Hydrophobia, and you will, I assure you, be instrumental in saving many of your fellow beings from misery and an untimely grave."

ANSWER.—The best known natural prevention of the disease called "Hydrophobia" is, a wholesale destruction of all dogs in Christendom. But the pound of "cure" is more popular. Of the symptoms of *Rabies* in the human being we need not say anything. They are, unhappily, familiar to thousands. The poison may remain latent in the human body for days, weeks, months, and even *years*, and then break forth with all its terrible symptoms of paroxysmal destruction.

TREATMENT.—The wound just made by the bite of a rabid animal, should be thoroughly and instantly cut open, so that the vessels will bleed freely, and all the parts should be immediately syringed or soaked with diluted aqua ammonia (spirits of hartshorn.) This remedy will give much pain, but it is of utmost importance. It is a good precaution to bind the upper portion of the wounded limb firmly with a strong ligature, in order to prevent the absorption of the subtile vapor into the circulatory system. This bandage may be safely removed in a few hours. After thoroughly drenching the wound with weak spirits of hartshorn (spirits of camphor will sometimes act as a substitute,) the parts may be drawn together, and then carefully covered with *arnica* court-plaster. This plaster can be obtained of any homeopathic physician, and of most druggists. (Better get the ammonia and the plaster at once, for 'tis said that "a wise man foreseeth an evil and hideth himself.") One thing is necessary—*i. e.*, a free discharge of venous blood from the parts bitten. If there is tardiness in this respect, apply a suction force to the wound without delay. All this will act as a preventive.

REMEDY IN LAST STAGES.—In case the system is contaminated with the *Rabies*, and the premonitory symptoms of the culmination begin to appear, then have nothing whatever to do with drugs. Tinctures of scull cap, musk, elecampane, chick weed, &c., &c., *ad infinitum*, are worse than nothing ; let calomel, and all the myriad forms of recommended drugs, entirely alone ; and instead, put the sufferer into a very hot vapor bath, as near 150° Fahrenheit as possible, and continue the sweating process for nearly twenty minutes. After which astonish the patient with a shower of cold water, and immediately cover him with blankets in bed. This will be followed by a desire to drink some cold fluid. Instead, give a tumblerful of strong red-pepper tea containing one tea-spoonful tincture lobelia. (Procure a bottle of this tincture *now*.) Perhaps a paroxysm will immediately come on ; if so, repeat the sweating process, with even more heat, if possible, for ten minutes ; then, as before, give the tumbler of cayenne tea and the tincture of lobelia.

It may be necessary to put the patient four times through this terrible ordeal. But as soon as the system is excessively weakened by the steam process, and the sickness at the stomach is succeeded with copious vomiting, the crisis is passed, and all danger of a fatal termination is removed. While vomiting, it is well to give a little spearmint or sage tea. If, however, the paroxysmal indications reappear, let nothing deter you from administering, still more vigorously, complete repetitions of the course above prescribed.

In some incorrigible temperaments the *virus* of the animal *Rabies* resists the most energetic remedies for several hours, and even days, especially after the final symptoms have been fully developed. The *snake-stone* of most countries, and the *mad-stone*, also, will neutralize the psychological vapor of *Rabies*, but not unless the individual is aware that the charm is within his possession. For it should be borne in mind that one-half of the deaths by hydrophobia, as by cholera, owe their origin to the erroneous belief of the patient that death is inevitable.

Vapor Baths for Hydrophobia.

Some twenty years ago the following case was stated in the Boston *Transcript :* A gentleman suffering from hydrophobia conceived the idea of suicide by means of a hot vapor bath. He entered the room when the temperature was about two hundred degrees, expecting to be suffocated, but soon fell into a profuse perspiration, and was permanently cured. Sometimes a sudden plunge in cold water will break the paroxysm

Galen's Remedy for Hydrophobia.

A favorite remedy for the bites of the rattlesnake, and in cases of hydrophobia, was introduced by Galen, under the name of "Mad-wort." The Latin name for it is "*Marrubium Vulgare*," which, being boiled down and done into plain English, is the common "Hoarhound," so generally and wisely used in cases of coughs and asthmatic affections. There is, unquestionably, a medicinal quality in this "mad-wort," which, if persistently used both in the form of poultices and as a drink, would neutralize the venomousness of snakes and rabid dogs. Let this be remembered when no other remedy is at hand. There is much experience in its favor.

Coup de Soleil, or Sun-stroke.

This malady is produced by exposure to the heat of the sun's rays. It is a modified form of apoplexy, and is most likely to attack persons of bad digestion, or of bilious and intemperate habits. The brain fever which succeeds a sun-stroke, is dangerous. Prevention is better than cure. Experience has shown [says a military writer,] that troops serving in warm climates greatly need protection from sun-stroke, often quite as dangerous and fatal as the fire of the enemy.

The judicious care and foresight of the British officers, in the recent campaigns in India and the Crimea, protected their soldiers from the danger, by *thick white linen cap covers*, having a cape protecting *the back of the neck*, which reflected, instead of absorbing, the heat of the sun.

Farmers and teamsters—indeed every one much exposed to the sun's rays in mid-summer—might save themselves from headaches and sun-strokes by this simple cap. Sun-strokes may also be prevented by keeping a wet handkerchief or napkin in the hat. This is the best remedy for any person whose head is accustomed to ache while exposed to the sun's rays. Always have two or three small holes in the top of your hat.

Proper Amount of Sleep.

Sleep is of the utmost consequence to the nervous and debilitated. But it should be had at night or before dinner; not afternoon, nor at irregular periods. Henry Ward Beecher hath well said : Men vary with regard to the need of sleep. A nervous man can get along with, perhaps, from five to six hours' sleep, while, perhaps, a phlegmatic man requires to sleep from eight to nine hours. The amount of sleep which a man requires depends upon his temperament. It seems strange to some that the most active men sleep the least. Men that work fastest sleep fastest. A nervous man does everything quick; he sees quick, and hears quick, and steps quick, and works quick, and sleeps quick. He does twice as much in an hour as a phlegmatic man, and he only requires half the time in which to do up his sleep-work that the phlegmatic man does. Every man ought, from his own experience, or from the advice of a physician—one who knows something—to determine what amount of sleep he needs, and then take that amount. He that steals necessary sleep from the night, steals from the Lord. He commits a theft for which God will visit him with punishment, in the shape of suffering and premature old age.

Sleeplessness—its Cause and Cure.

We know a person who presents a wonderful instance of wakefulness. No medicines can cause him to sleep naturally ; not even drowsiness comes upon his diseased and tremulous nerves; and he is moved in despair to exclaim :

"Oh. Sleep:
Nature's soft nurse ! how have I frighted thee,
That thou no more wilt weigh my eyelids down,
And steep my senses in forgetfulness ?"

Cause and Remedy.—The cause of all sleeplessness is inseparable from the condition of the ganglionic system. The centers of life are temporarily changed, and the mind trembles and wabbles in its orbit, somewhat as a planet would if thrown from its polar relations to the sun. *The Remedy*, therefore, cannot be found in anodynes, opiates, &c., but in the re-establishment of the nervous and cerebral motions. This can be accomplished by the "Movement Cure" in some cases; in others, the result may come on by applying the magnetism of a harmonious and healthy operator. Chronic wakefulness may be greatly controlled by eating plentifully of *onion soup* from twice to thrice per week. All sedatives, anodynes, and somnolent medicines, —such as opium, morphine, laudanum, hasheesh, tobacco, &c.— are at variance with the natural repose of the sympathetic ganglia. The loving and harmonious are invariably the *sweetest sleepers;* but the stoutest slumberers are they who work much and think little.

How to Procure Refreshing Sleep.

An unbalanced condition of the brain and its dependencies constitutes the cause of an "oppressive fullness" in many cases, and also explains the source of much mental fatigue and bodily weariness, on awakening in the morning.

Remedy.—Every night, before retiring, subject the posterior portions of your head to a thorough rubbing, chafing, and smiting, by the hands of another. While endeavoring to sleep, do not forget to breathe fully many times. The habit of reading and thinking, or of talking upon any exciting topic, after nightfall, is never promotive of the physical harmonies. Extreme sensitiveness, either physical or mental, should be regarded as a morbid condition.

Bad Dreams every Night.

We know a truly refined and poetry-loving person who is troubled nightly with dreams, the most grotesque and detestable. The case is not unusual.

REMEDY —Do not gratify your appetite with too many kinds of food. When a child, you were injured with affectionate expressions in the shape of *candy, raisins, nuts,* and *rich cake.* Yes, you were, good patient—don't deny it. These are the worm-generating "evil spirits" that now beset you in dreams. Sleep with your head toward the North Pole hereafter, and always go to sleep on your right side. Eat or drink nothing after seven o'clock P. M.

Origin of Physical Beauty.

A family physican (who is a sort of Abernethy in his way,) says: "You may laugh at me as you like, Miss, but, I tell you, it is a positive fact, which you are at liberty to disprove if you can, that, when Venus rose from the sea, the rising took place the very first thing in the morning, or else she never would have been the beauty she was!"

Earth's Polarities at Night.

In this world, the external body is identified inseparably from the spiritual body; that is, the Motion, Life, and Sensation of the latter are distributed throughout the former, as water through a sponge; so that, whether in sickness or in health, the two bodies are practically and consciously one and the same. The distinction between these dissimilar bodies is realized in advance of death only by those who enter the trance, or the state of complete clairvoyance.

All the earth's electrical currents flow northward. These are cold. The human brain is supplied with more blood than all the body besides. The head is, therefore, warm and magnetic, but sleeping with your head northward is best, because then the earth's cold fluids flow over it, and preserve a healthy state. The feet will be warmer in proportion. While sleeping with the head southward has the effect to deprive the extremities of their proper animation, and to fill the brain with extra heat and uneasy dreams. East and North are healthy; while South and West will produce nervous disease in the sleeper; but these remarks apply only to this latitude and longitude.

The Gospel of Self-Healing.

As all disease is the opposite of Health, and as all discord is the opposite of Harmony, so is the power of Health and Harmony identical with that by which disease is maintained. We would, therefore, counsel all fathers and mothers to heal their children by "laying on of hands;" not in anger, but with the full knowledge that healthful magnetism is the *natural medicine* for all sick bodies. The nerves convey disease or health in proportion as they are falsely or righteously influenced.

Unequal Bodily Development.

A correspondent writes a historical statement of his daughter's physical condition, which reveals a sad case of unequal development of the limbs, to which no ordinary treatment is applicable; but it may be possible to bring the motor nerves of such a patient under the control of the will of Indian spirits, in which case there are great probabilities of cure. Under ordinary circumstances, however, we counsel a castor-oil rubbing over the affected parts, every other night, with rapid manipulations on the neck and across the hips. Do not despair, for such a patient may become a light in your home! We would not have you administer any drugopathical preparations.

Confidence in Mother Nature.

QUESTION.—" Why is it that poets, philosophers, and men of science, as a general thing, discard *revealed* religion ?"

ANSWER.—Minds that think independently and with becoming care—expanding day by day with the out-flowings of facts, beauties, laws, and principles—very soon ascertain that God made Reason, and, therefore, *did not make an unreasonable revelation for man's guidance!* God's natural revelation *is* congenial alike to the fool and philosopher, to the Hindoo and Christian, while man's written revealments *are suggestive and spiritual,* yet invariably egotistic and uncertain. Wordsworth says :

> " Nature never did betray
> The heart that loved her ; 'tis her privilege,
> Through all the years of this our life, to lead
> From joy to joy; for she can so inform
> The mind that is within us, so impress
> With quietness and beauty, and so feed
> With lofty thoughts, that neither evil tongues,
> Rash judgments, nor the sneers of selfish men,
> Nor greetings where no kindness is, nor all
> The dreary intercourse of daily life,
> Shall e'er prevail against us, or disturb
> Our cheerful faith, that all which we behold
> Are full of blessings."

Put confidence in your nature, and you shall be strengthened. The gods of the beautiful Summer World will bless thee. By daily obedience to the *Self-Healing* principles of life, thy condition will become more attractive to the upper good. The Eternal Father and Mother of "spirits" have written their commandments in the human constitution. Look within and upward. Read such books only as contribute to the sum of practical wisdom. Do not strive to acquire knowledge too rapidly—think, act, enjoy.

Remain in your own Climate.

The climate of Pisa, in Italy, has been long considered favorable to persons afflicted with bronchitis and consumptive diathesis. It is very mild and moist, but many times relaxing and oppressive, to persons of northern nationalities. We would recommend you to remain in your own climate, obey the laws of Nature, keep your spirit happy by doing good deeds, and take those simple remedies prescribed in this book for consumptive conditions.

> "Oh Health!
> Oh happiness! our being's end and aim;
> Good, pleasure, ease, content, whate'er thy name,
> That something, still, which prompts the eternal sigh,
> For which we bear to live, or dare to die."

Softening of the Brain.

This condition is very common; the heart is proportionally *hardened*. Professional, but more frequently business men, are its subjects. The predisposing cause is sumptuous living. After a morning fully occupied with business matters, a man comes regularly to a dinner of various and highly-seasoned dishes of fish, and fowl, and flesh, with every adjunct to excite and gratify the appetite. He partakes freely of food and wine, in excess, to be sure, though, perhaps, never to the extent of gluttony or inebriety. The papers are read, cigars are smoked, a few hours are passed socially, and the evening closes with a hot supper and abundant punch. If a man living thus continues successful in his plans and his business, he may go through life with no other physical or mental infirmity than the pain and irascibility of gout, or the distress and gloom of dyspepsia. But if it be otherwise, if he meet with a reverse of fortune, or if some grief or chagrin come upon him, then he is exceedingly liable to this fatal disease, which is the joint product of luxurious living and some torturing anxiety or disappointment.—*See Report of* Dr. John E. Tyler, *Superintendent of the McLean Asylum.*

How the Will Acts on Nerve-Centers.

Physiologists cannot determine in the living subject the exact condition of the nerve-batteries located at the base of the brain and in the spinal cord. These centers in color are gray, derived from the positive substance of the cord; and by fibers (or thread conductors,) they communicate with every muscular tissue in the organization. Your Will can, through the magnetic forces of these centralized batteries, reach and ramify through every part. Practice a few days, as we have admonished you, and your experience will sustain our philosophy.

One correspondent writes that the Willing-remedy, applied to herself for deafness, has already benefited the parts; but complains that the exertion of Willing, accompanied with the suitable manipulations, has the effect to induce sleep.

Many patients would be delighted to have similar results follow their pneumogastrical efforts. If, however, every patient so affected would take a foot-bath of tepid water, while in the act of manipulating and Willing, we are sure that a cheerful wakefulness and other benefits will ensue.

The Will-energy at Work.

See how that fellow works! (says a writer); no obstacle is too great for him to surmount; no ocean too wide for him to leap; no mountain too high for him to scale. He will make a stir in the world, and no mistake. Such are the men who build our railroads, dig up the mountains in California, and enrich the world. There is nothing gained by idleness and sloth. This is a world of action, and to make money, gain a reputation, and exert a happy influence, men must be active, persevering, and energetic. They must not quail at shadows, run from lions, or attempt to dodge the lightning. *Go forward zealously in whatever you undertake*, and we will risk you anywhere and through life.

The Man who can Will.

Mirabeau said: " Why should we feel ourselves to be men unless it be to succeed in everything, everywhere. You must say of nothing, *That is beneath me*, nor feel that anything can be out of your power. Nothing is impossible to the man who can Will. *Is that necessary? That shall be :* this is the only law of success." Whoever said it, this is in the right key. But this is not the tone and genius of the men in the street. In the streets we grow cynical. The men we meet are coarse and torpid. The finest wits have their sediment. What quantities of fribbles, paupers, invalids, epicures, antiquaries politicians, thieves, and triflers, of both sexes, might be advantageously spared! Mankind divides itself into two classes—benefactors and malefactors. The second class is vast, the first a handful. A person seldom falls sick, but the bystanders are animated with a faint hope that he will die: quantities of poor lives; of distressing invalids; of cases for a gun. Franklin said : " Mankind are very superficial and dastardly; they begin upon a thing, but meeting with a difficulty, they fly from it discouraged; but they have capacities, if they would employ them." Shall we, then, judge a country by the majority, or by the minority ? By the minority, surely. 'Tis pedantry to estimate nations by the census, or by square miles of land, or other than by their importance to the mind of the time.—*Emerson.*

Gall and Spurzheim's Works.

In the history of the investigations of phrenological science, we first meet with the celebrated Gall. This physician was the first to make practical observations upon the living brain. But Doctor Spurzheim's classifications and works have superseded those of Gall in popular estimation, because the former was the most successful in bringing the facts of mind more clearly and simply before the world.

It has been ascertained that the front lobe of the brain exerts an influence which is invigorating and refining to all the senses and the nervous system, but that its effect upon the muscular and osseous systems is somnolent and debilitating. Harmony is possible only when both brains, back and front, are equally exercised. (See "A Remedy for an Unbalanced Body and Brain," on another page.)

What the Will can Do.

QUESTION.—"Do you mean to convey the idea that a person who has inherited *nervous* infirmities may ever, by careful, conscientious use of the internal, self-healing powers, become so 'redeemed and sanctified' as to have 'a sound mind in a sound body'? Can one with undeveloped concentrativeness make use of the practice recommended in Pneumogastrical remedies?"

ANSWER.—Yes; we certainly mean to teach that mind is destined to stand sovereign master over all *below* its exalted plane. Matter—the body—is below the soul; therefore *the soul* is capable of instituting an absolute government. "Practice makes perfect," is an old and a true proverb. The method of applying the Will-power is as simple as walking, moving your hand, or speaking, for all such motions are from the Will. Why not extend the operations of this power over all parts of the dependent organism?

Forms of Temporary Insanity.

A Friend of Reform wants to know what medicines or preparations will produce temporary insanity—meaning a mental derangement of a week or two—and then pass off without entailing disastrous consequences.

Answer.—There are several medicines, besides alcohol, capable of inducing an insane condition of the brain. Our Insane Asylums show that from one-fourth to one-third of all cases admitted, have been made insane by the habitual use of alcohol. This fluid, although it mixes readily with the serum (or water) of the blood, never ceases to be alcohol. It produces a contraction and condensation of the tissues, and liberates the brain for the time being, very much to the enjoyment of the mind and social feelings; then ensues the second stage, called inebriation, or drunkenness, which is an insanity, exhibiting a melancholy derangement of the intellectual faculties; and lastly, the individual is correspondingly *depressed*, relaxed in all the fibers, and rendered unfit for the manifestation of either mind or muscle. The middle stage is productive of all those oddities and eccentricities which usually characterize the inebriated individual. Thus, by the temporary insanities produced by alcohol, we get:

1. The Fighting Drunkard.
2. The Social Drunkard.
3. The Mirthful Drunkard.
4. The Political Drunkard.
5. The Burly Drunkard.
6. The Cowardly Drunkard.
7. The Melancholy Drunkard.
8. The Religious Drunkard.
9. The Blasphemous Drunkard.
10. The Voluptuous Drunkard.
11. The Sentimental Drunkard.
12. The Beastly Drunkard.

You will observe that, fundamentally considered, the condition of the Drunkard is that of insanity. The variations are traceable to the natural propensities of the individual character, which, at such times, are not modified and controlled by the deeper life and understanding. This sort of mental derangement may be produced by the administration of a great variety of preparations. American Hellebore (*Veratrum Viride,*) Henbane (*Hyoscyamus,*) Foxglove (*Digitalis,*) Skunk's Cabbage (*Symplocarpus,*) Thorn-apple (*Stramonium,*) Indian Hemp (*Cannabis Indicus,*) Tobacco, Prussic Acid, Spurred Rye (*Ergot,*) &c., &c., and yet many other sedatives and inebriants may be mentioned, which, given either in tincture or by decoction, will, if taken in continued or over-doses, produce temporary insanity, and many visionary symptoms, always varying with the hereditary characteristics of the individual patient. The consequences of temporary insanity are not necessarily lasting or disastrous. And yet, years are sometimes wasted in the effort to restore diseased nerves.

Brain Rest Absolutely Necessary.

Brain rest is sometimes better than medicine, particularly in cases where the nervous system is deranged, and the digestion enfeebled. We cannot promise restoration of a nervous patient, even under favorable conditions, in a period of less than six months. We counsel you to engage in some employment not intellectual, during the period consecrated to the process of recuperation. Over you, at all times, is the Summer Land. You will receive, if you deserve, aid and comfort from its inhabitants.

Cause of Brain Fits.

In some cases we detect a diseased condition of the *ganglia*, extending each side of the spinal column. The upper portion of the spinal cord, the *medulla oblongata*, is enlarged as by an inflammation in its substance. The *motor nerves* are, consequently, disturbed; the right side being sometimes more diseased than the left, imparting an involuntary wish and tendency to roll or whirl as an amusement. As the nerves of seeing, and hearing, and tasting, arise near the diseased parts, it is but reasonable to expect that these senses will be gradually impaired.

Remedy.—There are but two paths of approach to the seat of the disorder—one, through the nerves of the stomach; the other, through the spinal column. Raw onions should be applied externally over the stomach every day, in the form of a poultice; also a small one, well macerated or grated, should be administered every morning and evening, mixed in molasses if desired. Magnetic treatment is capable of reaching the throne of the disturbance, through the spinal column. It should be used faithfully, with the human hand. It is necessary to keep the patient's head erect, or in a natural position, which may be effected by girdling the neck with a hair cushion.

Treatment for Epileptic. Fits.

Fits of unconsciousness are caused by a sudden strangulation in the nervous circulation, which, from many causes, may take place along the track of the Pneumogastric nerves, somewhere between the brain and the bottom of the stomach. An account of symptoms is deemed unnecessary.

Remedy.—The philosophy of curing this frightful malady is simply the removal of the tendency to arterial strangulation, and the consequent nervous suspension in the route of the sympathetic and pneumogastric nerves. An equal distribution of the vitalic forces is absolutely necessary. We, therefore, counsel you never to yield to the temptations of your appetite, to over-eat or to drink largely of any fluid. When a fit of hunger overtakes you, beware! for even then *inverted* Nature is preparing to adjust itself by a shock called "epilepsy." This surprise of the nervous system is accomplished by the righteous operation of long unbalanced forces. But, unhappily, if Nature does not attain her balance by a few shocks, the tendency is toward a greater loss of equilibrium in the brain and its ramified influences. Therefore, a gradual failure of memory and of intellectual vigor, even to imbecility, are the resulting consequences.

We make these remarks in order to enforce the importance on parents, and mankind generally, of early attention to the fundamental causes of this malady. Remove the nervous strangulation, and instantly all the symptoms will depart; and the same is philosophically true of all nerve-pain, cramp, spasms, and paroxysmal affections. But in advanced stages of this disease, the treatment must be varied to suit temperaments and occupations. In every case of long-standing, we prescribe butter milk, whenever the thirsty symptoms begin, and it may be drank plentifully in response to the extra sensations of hunger. Wear a compress or bandage of either crushed wormwood or life-everlasting around the waist every night. During the day wear a little linen sack of equal parts of powdered belladonna and iron-rust directly on the pit of the stomach. Do not fear the progress of this disease after you adopt the means of overcoming the causes in the arterial and nervous systems.

Fits of Indigestion.

Sometimes what is termed "Falling Sickness" is caused by a peculiar kind of indigestion. In such case, the disease becomes sympathetic with the digestive functions, and may be controlled mainly through the pneumogastric nerves.

REMEDY.—Never eat out of season; nor between meals; **not**

even an apple or a few nuts. No fruit or berries for supper. A rigid system of hygiene must be enforced by the authority of your will and reason. Make the following paste: Powdered leaves of stramonium, one ounce; flower of sulphur, two ounces; Spanish fly (*cantharides*,) powdered, twenty grains; iron filings, one ounce; and a small piece of burgundy pitch. Amalgamate these ingredients over a slow fire by stirring constantly. Put about one-quarter of this *magnetic paste* into a thin kid sack, and wear it next to your body, immediately over the pit of the stomach. It should be renewed about once in every six weeks. This, in connection with your Will-power and correct habits, will cure.

More than half of your distressing infirmity is caused by a periodical stagnation of nutrition in the small intestines. The chyle is not appropriately absorbed through the mesenteric system. The sympathetic and pneumogastric nerves enter their just complaint at the throne, and seek from the brain (and the mind,) a full recognition of their wrongs, and plead for ample reparation. The effect of this struggle is a prostrating paroxysm.

Brain Fits and Incipient Epilepsy.

"Some two or three hours after eating (says a correspondent,) frequently, my eyes begin to blur, and after awhile I become quite deranged, very nervous, and at times so nervously excited that it would take ten men to hold me. These paroxysms will last about an hour, when I am *suddenly* relieved, and become exhausted and weak, perspiration will roll off me, and my nervous system is all unstrung. The only relief I have ever found has been obtained by eating! Eating will stop the trouble, if done when my eyes begin to blur; though lately that has not had so good an effect as formerly. I have increased very much in flesh within a few years."

REMEDY.—The case is similar to one we treated and cured in 1847. Memory brings back every internal fact of the patient.

There was a disease all through the mesenteric system. The lacteal vessels refused to drink up the chyle at the right hour; the sympathetic nerves promptly reported the rebellion at the brain; the brain sent dispatches over the pneumogastric nerves that the digestive collections *must go* forward; and so energetically did Nature *insist and persist*, that the patient was thrown into violent paroxysms—*incipient epilepsy*. Our treatment was very simple, and the same is adapted to many persons —viz: Take the extract of chamomile, half a drachm; camphor, ten grains; cayenne pepper, one scruple. Of these substances, well mixed, make seventy pills. Take two about an hour after breakfast and dinner for one week. Subsequently take but one, and only when the symptoms re-appear. Use the Will-power energetically.

Philosophy of Neuralgia, or Nerve-pain.

Very many suffer from periodical attacks of this indescribable pain. Sometimes the distress fixes upon nerves within the stomach, in the left side, and often the pain is most unendurable in the forehead, behind the eyes, showing that there is an exact correspondence between different sections of the lower organism and different strata of the brain. Thus: If the bowels are diseased, the base of the brain is disturbed, and the patient suffers alternately in both regions; in like manner, the middle portions of the brain sympathize with all derangements of the liver and stomach; and in consumption, as in all cases of bronchial irritation and debility, the superior parts of the brain are involved in more or less disturbances. For this cause. as general observation declares, all consumptives are more hopeful and less desponding during the several stages of the malady than patients afflicted with any other known disease. This is true, because the superior organs of the brain are affected and stimulated to activity whenever the upper part of the lungs are diseased.

With these explanatory remarks, we call attention to the causes of nerve-pain, and affirm that no human being can experience any such pain unless there be first a compression or embarrassment somewhere in the functional or circulatory system, by which the blood has been forced into some capillary or hair-like vessels, thus tying up, or *ligaturing*, so to say, some of the important *nerves* which convey the elements of vital-life from the brain-batteries to the different organs in the bodily structure. And, therefore, no one can ever be perfectly cured of what is termed "neuralgia," until the circulatory systems are in perfect running order. Not a particle of bile, of broken-down blood, or of unworthy material, must float in the empire of veins and arteries. All magnetic-passes, all palliative mixtures, all nursing, and petting, and waiting "patiently" for the pain to subside, will amount to almost nothing. If Nature is able to restore the balance of circulation, with or without medicines, the patient is *suddenly* relieved and cured; but if the internal *ligaturing* of nerves by the blood-vessels is established, then, alas! the victim is destined to be a sufferer of pains, more or less endurable and aggravating, for months or years, or until the angels unlock "life's flower-encircled door," to show the worn-out pilgrim of earth the holy scenes of the Summer Land.

REMEDY.—Many plans may be adopted to restore the arterial and venous circulation to a perfect balance. The Thompsonians and Hydropathists are about equally successful in accomplishing this result. They sometimes cure severe cases, as do the Homeopathists also, with a marvellous degree of celerity and completeness; but we know of instances of incorrigible *nerve-pain*, wherein the chieftains of every system of medicine, and even Nature herself, gave the patient over to the law of Progression, which steadily marches through the tomb. The law of Cure, however, is as plain as the eleventh commandment, viz: Establish an equality of operations between part and part. That is, equalize the bodily temperature; distribute the activities of the heart, and brain, and stomach; and then, having liberated the *nerves* from their special embarrassments, the patient will experience instant relief; and, with care, the sufferer may never have a return of the indescribable agonies of neuralgia.

Let all patients, suffering with this nerve-pain, commence at once to take two thorough bodily sweatings every day, either by means of vapor baths, or by packing in many folds of cotton and woolen quilts. It will be necessary to drink some very warming tea, such as Crawley, or Ginger and Red-pepper, in order to facilitate the process of perspiration. The entire body should be bathed with bay-rum, or with weak vinegar and salt, immediately after the sweating, which should continue at least fifteen minutes. The patient should invariably wear sheepskin moccasins at night, and during the day also, if the pain threatens to return.

Drug Treatment for Neuralgia.

Some time since we published a recipe to cure neuralgia. Half a drachm of sal ammonia in an ounce of camphor water, to be taken a tea-spoonful at a time, and the dose repeated several times, at intervals of five minutes, if the pain be not relieved at once. Half a dozen different persons have since tried the recipe, and, in every case, an immediate cure was effected. In one, the sufferer, a lady, had been subjected to acute pains for more than a week, and her physician was unable to alleviate her sufferings, when a solution of sal ammonia in camphor water relieved her in a few minutes.—*Med. Journal.*

Remedy for a Sudden Neuralgia.

In cases of violent attacks of nerve-pain in neck, teeth, face, ears, and head, we recommend the following mixture; Wine of opium, 30 drops; sulphuric ether, 1 scruple; fluid extracts of belladonna and yellow jessamine (*i, e., gelseminum,*) of each 1 drachm; put into 1 ounce of lavender water. Saturate a linen cloth with this mixture, and apply it for an hour or more over the region of the pain.

Spasms, Periodical.

In such cases there is much irritation in the *medulla oblongata*—meaning that portion of the spinal cord which is situated within the bones of the skull—and the tendency is to the rapid development of tubercles. The patient is subject to spasmodic fits about every ten days, preceded by a craving appetite, hot head, cold extremities, and sallow complexion. The patient may be fleshy, and well developed in physical proportions, and yet in general health be very unsound.

REMEDY.—Watch the approach of the premonitory symptoms. Face them resolutely, and work to bring about a *balance of temperature*. Wrap the feet and legs in many folds of flannel. If they do not freely perspire, but are only hot on the surface, then chafe them with a little flower of mustard before enveloping with flannel. Stop all food of every kind, except a little porridge, full twenty-four hours before the spasms usually appear. During the crisis, let the patient rest on a small pillow filled with hops and a little camphor.

A Passive Use of the Pneumogastric Remedy.

MR. DAVIS: Allow me to give a testimony concerning that which has proved of much benefit to myself, and might be of use to others. I tried your first general directions for the applications of the Will-power (given in your HERALD OF PROGRESS, of June, 1860,) but without receiving any benefit. Being susceptible to influences of spirit-mind, I got the impression that I should assume the passive instead of the positive condition, and become as negative as possible, observing the same rules of time, &c., as you directed. I soon felt the recuperative powers of my system becoming active. Six months have passed, and I have been gradually, but steadily, gaining all that time.

I reason of the change in this wise: Mine being an active positive temperament, the Will-power had exhausted or overtaxed those faculties or organs of my system which were necessary to convey the life principle to the diseased parts; consequently, passivity, rest, and the influx of spirit-power which I invoked at the time, and very perceptibly felt, was best adapted to my condition. Thine, for truth and investigation,

MARY M. BISHOP.

There is a Purpose in Pain.

We have, from the first, described "pain" as the language of discord in the soul-principle; it is the voice of divine life in man, crying for health and happiness. It seems that Owen Meredith does not teach differently. He says:

> "There is purpose in pain;
> Otherwise it were devilish. I trust in my soul
> That the Great Master Hand which sweeps over the whole
> Of this deep harp of life, if at moments it stretch
> To shrill tension some one wailing nerve, means to fetch
> Its response the truest, most stringent, and smart,
> Its pathos the purest, from out the wrung heart,
> Whose faculties, flaccid it may be, if less
> Sharply strung, sharply smitten, had fail'd to express
> Just the one note the great final harmony needs.
> And what best proves there's life in a heart?—that it bleeds!
> Grant a cause to remove, grant an end to attain,
> Grant both to be just, and what mercy in pain!
> Cease the sin with the sorrow! See morning begin!
> Pain must burn itself out if not fuel'd by sin.
> There is hope in yon hill-tops, and love in yon light,
> Let hate and despondency die with the night!"

Remedy for Pain in the Face and Neck.

In reply to the questions of many correspondents, we republish the following: Extract of cicuta and stramonium, each five grains; laudanum, half an ounce; brandy, two ounces;

Put these ingredients together, and add one ounce of alcohol; then warm and shake the composition until its constituents are in a complete state of amalgamation. (Get the medicine and keep it in the house.)

DIRECTIONS.—Bathe under and behind the ear, under the jaw, on the cheek over the aching teeth, and put a piece of cotton, saturated thoroughly with the liquid, into the ear; change this frequently, continue to bathe as directed, and relief will almost immediately follow.

Pain in the Neck of Housekeepers.

There is a wonderful network of glands, batteries, and life-conductors, or nerves, in the back parts and at the sides of your neck, which are disturbed by clogging matters whenever you have over-exercised about the house. The true remedy consists in bathing your neck and back-head thoroughly every morning, in pure cold water; to be followed by self-manipulations and gymnastic exercises. Squeeze your throat whenever it is threatened with soreness, and gargle with red-pepper. Chew a few chamomile flowers before breakfast.

Remedy for Pain in the Joints.

The causes of such pains and lameness are entirely "too numerous to mention." We will tell you how to remove every pain and ache, which originates from rheumatic and negative conditions in the joints, or from a sudden sprain and local tension of the tendons near a socket. Make a powerful decoction of burnt coffee, say half a pint; to which add four ounces of chloroform, and one ounce of cajaput oil. This magnetic wash should be kept tightly bottled, and used very sparingly; a few drops at a time on the parts affected, to be followed immediately with manipulations for ten or fifteen minutes.

Sciatica Neuralgia.

This is a disease of the sacro-sciatic nerve. The pain, lameness, and special symptoms, may exist independently of other nervous derangements. We have examined numerous cases of this kind, but never have found what some physicians term an inflammation of the nerve; although, in many instances, the *neurilemma* (or nerve-sheet,) has been considerably enlarged by an excess of fibril heat. The nerve-pain is principally owing to a sort of strangulation of the nerve itself. The life-principles of health and vigor cannot freely circulate through their appropriate conductors. Sciatic pain, or rheumatic neuralgia, is a common effect from such conditions. There is no quicker or shorter road to take to reach healthy conditions, than that of *hand-friction* and magnetization. The parts need to be *rubbed* vigorously, and then manipulated, every day. When the pain is severe, apply a poultice of burnt salt and baking soda—one teaspoonful of the latter to four table-spoonfuls of salt—mixed in a sufficient quantity of common meal.

Pain Between the Shoulders.

Your excruciating shoulder-pain is caused by a derangement of the hepatic functions. The pneumogastric nerves report your liver disturbances in the shape of pain, and a fatiguing ache on the right shoulder, near the neck. The nerves of the head sympathize and throb with the confusion of sensations in the system.

Remedy.—Adopt the Will-cure at once. Straighten the body, and breathe deeply several times in the fresh air before breakfast, which should consist only of one boiled egg and a piece of stale bread. Not more than one cup of any warm drink. Practice throwing your arms backward, and, meantime, give your shoulders a variety of sidewise and churning motions.

Indigestion is the main cause of the whole disturbance. Let some good friend *pound* and pommel you about the waist before going to bed.

Pain in the Right Side.

As a general remedy for right-side pains, we recommend a thorough manipulation of the left side, from the shoulder down to the hips, so as to affect the action of the spleen, kidney, heart, stomach, and intestines. These organs and parts, by being invigorated and surcharged with functional activities, will send aid and health to the suffering on the opposite side. This is the best method, often, when one side is healthy and the other diseased. Surcharge the healthiest parts, and they will naturally magnetize their disabled neighbors in the visceral cavity.

Sudden Pain in the Breast.

This peculiar pain is caused by a sudden compression of small vessels, accompanied with the generation and confinement of air in the cellular tissues, all which indicates a weakness in your general system. The best treatment for immediate relief is a thorough steam-bath; keep up the sweating for many hours, by means of hot bottles, &c., and the use of crawley tea. Wear a very warm garment or bandage of lamb's skin next to your body in winter. Frequent sweating will aid you; always avoiding sudden changes: and exercise in the house when you cannot go out.

Deficient in the Cords of the Leg.

A mother states that her daughter, ten years of age, is deficient in the cords of one leg. The muscle is not fully developed, the foot turns in a little, and there is every indication that the tendons, toward the extremities of the muscles of the limb, are deficient and nearly useless.

REMEDY.—It should be remembered that the muscles and tendons are like so many ropes and cords, by which parts of the body are changed or kept in proper position; and, further, that every muscle is compounded of many *fibers*, very delicate in material and structure, which are fed by the brain-life through the nervous system. The true remedy for a deficient muscle, therefore, is to be found in the brain of its possessor. Tell the patient, then, to fix the mind on the parts which are to be restored, and next, to throw a volume of WILL into the cords of the weak and trembling limb. Take each step with great and firm resolution, and stop walking the moment the resolution is weakened. To strengthen the general system, and to prevent spasms, we prescribe a bowl of wild cherry tree bark tea, for a few weeks, to be drank between Mondays and Wednesdays.

Hot Head and Feverish.

A friend writes: "My head is hot and feverish most of the time; my stomach is very much out of order; also my liver and kidneys are somewhat affected; in fact, I am diseased all over." REMEDY.—Use the Will-power, as heretofore described; correct your daily habits of eating; and take the following "rough and ready" remedy: Two gallons of old cider; add half a pound of rusty nails, and half an ounce of rhubarb-root. Let this preparation tincture for one week. DOSE.—Take one wine-glassful immediately on rising in the morning. Add one more gallon of cider in a few days. Eat plenty of stewed apples and bread for breakfast; no meat until dinner; never drink anything intoxicating.

Remedy for "Lock-jaw."

The terrors of *tetanus* may be dissipated by a strong, steady-nerved magnetizer. The disease is not necessarily dangerous.

It may be controlled by magnetizing the feet into a perspiration, and by administering the homeopathic preparation of *Aconite* and *Belladonna* alternated, or *Mezereum* and *Merciurius*, thirty minutes apart, for six hours. There are several methods of relaxing the contracted fibers, but we would place magnetic treatment before all the rest.

Remedy for a Stiff Ankle.

A patient writes an account of his stiff ankle, which has been disabled and troublesome for about three years. There is a hard swelling on the outside of the joint, as large as a hen's egg; and a kind of puff swelling on the inside; with much pain and wasting away of the limb. REMEDY.—Adopt the system of diet recommended for patients afflicted with itch or erysipelas. Frequently lay the hands on the parts affected, thus open the pores of the skin to let the disease escape, and keep up the operations every day until the swellings disappear. Apply nothing else during the day-time. Every night put your foot into essence of peppermint and sweet oil, bathe the swellings thoroughly, and then envelop the parts in cat's or woodchuck's fur. Use the entire skin with the fur next to the ankle.

Membranous Rheumatism.

There is a neuralgic disease of the membrane about the bones. This affection is known by exquisite sensitiveness of the limbs and joints, with occasional pain of the severest kind, but with little or no swelling of the parts and joints so suffering. REMEDY.—Take as much bodily exercise as possible without incurring excessive fatigue, and this every day. After such a walk, or (which is far better,) work out in the sun until quite warmed up, cover the affected parts with a preparation of soap, cayenne pepper, lobelia, and camphor. Thus: One pound of common soap; one tea-spoonful of red pepper; two ounces of fine-cut tobacco (or lobelia;) one table-spoonful of powdered camphor. Add water enough to blend these articles together over a slow fire. When cold, it should be of the consistency of common paste. Give the rheumatic parts a coating of this paste, and wrap them up, excluding the air, with thin leather or oiled silk. A great fact in all rheumatic affections, is inactivity of *portions* of the system. Bring all parts into action as rapidly as possible. Use such articles of diet as will keep the bowels alive every day. Avoid all narcotizing remedies. Opium, internally, is no friend in this disease; it gives you a ticket from Bad to Worse; better purchase a passport to the Highlands of Happiness. *Work your passage!* Examine the laws of health relative to diet, and do not dare to act contrary to your best light.

Remedy for Periodical Rheumatism.

Prepare the following magnetic liniment: Oil of hartshorn, of mace, and of lavender, of each four ounces; stir them into a prepared mixture composed of one pint of alcohol and the whites of six eggs perfectly amalgamated. On the first sensation of uneasiness of the rheumatism, give the parts a *severe* hand-chafing at night. It would be far better to get a friend to curry you down, and dress the affected members from top to bottom, as you would manipulate the limbs of a favorite trotting horse. Then apply the magnetic liniment in small quantities When the pain becomes severe, use the mixture twice or thrice per day; but it will be necessary at such periods *to sleep* in tightly-fitting flannel garments.

Periostitis and Headache.

There is more than one beloved patient who is strong-bodied, yet remarkably sensitive and weak-headed; not in the *intellect ual* part, but in the fibrous membranes, which clothe the surface of the bones. The head will ache, and the teeth also, even when the general health is sound. Shall we name it "Periostitis"? meaning a rheumatic inflammation of the *Periosteum*. Yes, that term is sufficiently descriptive of the seat and cause of frequent headaches. In such a case, put cold water behind your ears, and on the back of your head and neck, *before you bathe any other part of your body, each morning*. Do not wet the top and front of your head. Keep these portions dry in the morning.

Nervous Burnings and Pain.

It is not uncommon for an American woman to displace some of those physical energies that are necessary to the possession of harmony and happiness. She is excessively nervous at times. But anodynes and sedatives are not the remedy. For the painful and burning symptoms she should use no preparations in the shape of stimulating washes. Her pathway to Health leads through obedience to the pneumogastric and self-magnetic remedies. It will be necessary for the patient to retire very early, supperless, and without warm drinks of any kind. For breakfast and dinner, she may eat and drink whatever and as much as her appetite and reason will honestly prescribe. About two hours after dinner, each day, it will be necessary for friendly hands to rub and pommel the patient, as hard and rapidly as possible, without severe pain, upon the neck and shoulders, down the entire spine, and all the way around the waist. Meantime, remember, full respirations are necessary. Magnetism, alone, would amount to nothing to one in such a condition. The patient needs the benefits accruing mainly from vigorous mechanical pressure and judicious buffetings. But these latter remedies will also prove non-availing and injurious, unless bestowed by the hand of benevolence and sympathy.

Weakness and Emaciation.

A correspondent writes that his companion "is tall and slender, very poor in flesh, but rich in spirit; she has a lively, intelligent mind, quick of perception; and is as good a wife, and judicious a mother as can be found." But she is afflicted with a general incapacity of the muscular system. He says: "She is a strange compound of strength and weakness. She bathes with water every day; eats no animal food; works hard; can easily walk three miles per day; has a good digestion; good white teeth; regular bowels; regular and painless in her menstruation; warm feet, etc.; yet feels excessive weakness; faints sometimes; small appetite; dandruff on her head; is very emaciated and weary; slightly inflamed eyes; white, odorless discharge, in small quantities, from the vagina; is troubled with a sleepy feeling, but never with any such thing as depression of spirits, for she is a *Philosopher !*"

REMEDY.—There is a sure medicine for the emaciation and symptoms above described, which are not uncommon among women in this country. It will consist of dietetic influences for the invigoration of the absorbent and secretory systems. There is a deterioration in the constituents of the blood. There is no want of iron in the system (for the bile is sufficiently charged with ferruginous elements derived from milk, eggs, water, and vegetables,) and yet there is a want of appropriating power in the alimentary canal and symphatic absorbents generally. And to set the entire system in good working order, than which the music of the spheres may not be more harmonious, we recommend for breakfast, one soft boiled egg, with bread and butter, and one small glass of strong beer. With your dinner (which may be composed of anything you wish suitable for the season,) drink a tumbler of weak tea of wild cherry tree bark. For supper, which should not include fruit or cakes, you may drink a cup of Homeopathic chocolate. Experience has proved that a person thus afflicted, may eat all kinds of light bread and biscuit not containing potash, soda, or other similar ingredients, and not too fresh ; cakes composed of meal, eggs, sugar, and a little butter; buckwheat cakes not raised with fermenting powders ; light puddings and dumplings of wheat, Indian meal, rice, oatmeal or bread, without wines, spices, or rich sauces; hominy, Indian mush, rye mush, groats, pearl barley, potatoes, turnips, carrots, spinach, cabbage, cauliflower, asparagus, green or dried peas or beans. It will be necessary to use vaginal syringes, two or three times per week, of strong motherwort decoction, or warm water containing from three to five drops of sulphuric acid.

Remedy for Weakness and Pain.

This unbalanced condition between the sensory nerves and the nerves of motion, may be overcome by establishing an equal circulation of the magnetic energies between the brain and extremities. This equilibrium can be accomplished only by and through a persistent course of magnetic treatment. You should be pounded, rubbed, manipulated, and lastly, oiled from head to feet, at least once a day for many weeks. Home is the best place.

Treatment for Painless Paralysis.

Symptoms of paralysis in one or both limbs, affecting the face and brain, call for continuous magnetic treatment. The blessings of health will flow through man's form much quicker and more permanently under the proper external circumstances. But there is another method which may reach many sufferers, viz : Take one-third pound opium, macerate (soften,) it in sufficient water for two or three days, occasionally working it over with the hand. Then add muriate of ammonia, one ounce ; oil of cinnamon, or of cloves, two ounces ; alcohol, one pint. Mix thoroughly and add one pint of water   Dose.—A very little of this preparation will penetrate the magnetic membranes of the entire system. It must always be used externally, and in connection with olive oil in case the smarting is too severe. The spine should be bathed with this tincture once a day, from the neck downward—then your arms and legs—concluding by bathing the soles of the feet thoroughly. We pray that the watching angels will extend healing hands to every Brother.

Remedy for a Multitude of Sins.

A patient says : "About three years ago I got under what is called 'conviction for sin.' My feelings were horrible beyond the power of any pen to describe. Such feelings continued for six months, until my mind was almost lost and my body shattered. . . . Spiritualism has done wonders for me mentally, but the body is afflicted as follows : 1st. Nervousness. 2d. Pain in the region of the kidneys and across the small of back. 3d. Great prostration and general weakness. 4th. Fever most of the time, nearly always after dinner. 5th. Headache, gloominess of mind, disposition to look for evil. 6th. Costiveness, heat in head, face often flushed, pain in legs, sweat under the arms all the time, but not any other part of my body." This catalogue, though brief, covers the symptoms of very many of our countrywomen, and there are also diseased men, who have the most afflictive disturbances, including horrible feelings developed by "conviction of sin."

REMEDY.—Stop voluntary sinning from this hour, and your mental condition will improve wonderfully fast. As a medicine, eat nothing between breakfast and supper ; never eat any bread fresh baked ; no pancakes or salted food of any description ; with this understanding, we will prescribe for you—for everybody with these symptoms—the following Chylifier : Cinnamon bark and nutmegs, bruised, of each one table-spoonful ; cardamon seeds, horse-radish, ginger-root, mandrakes, and Turkey rhubarb, powdered, of each two table-spoonfuls ; let all these macerate—that is, soften and dissolve in water sufficient to cover them and more—for twelve days, in a light vessel, then strain and add two quarts of cider brandy or half a pint of alcohol. Use this remedy without sweetening, if you possibly can. DOSE.—A tea-spoonful, with or without water, at the usual dinner hour, and the same on retiring for the night. In some cases the dose may be doubled: not, however, if the feverishness is increased thereby. Take a warm water wash-off, or a thorough steam-bath, once every week, for a while.

Resignation as a Medicine.

The initiatory step toward a cure, in many nervous diseases, is *resignation*. Your body and all its nerves will improve when your mind *voluntarily* yields allegiance to existing conditions. Do so at once. Will yourself into a state of comparative carelessness. Become passive and resigned—not because of the necessity, but from wish and intention. It is a sad spectacle to behold a human soul *compelled* by disease to be passive and tranquil. Such is hopelessness and frail weakness. Perhaps this is not yet your condition. Listen, therefore, and be *strong* in your weakness. Resign virtuously and beautifully out of the still fountains of prayer and will. We style this course the first step to physiological restoration. We do not counsel you regarding food, sleeping, clothing, water, &c., for all these things you may have attended to from your youth upward. Only this: *Resign while you can*. Allow yourself to become an invalid. Have the luxury of doing while you *may* before yielding because you *must*. Observe, now, that we teach you to be passive spiritually, and thoughtless as to the action of medicines, and indifferent whether you get well or go hence. Yet you should exercise early every stormless morning. Return and resign to rest; the manipulations will benefit you. Let some friendly hand give your body a thorough castor-oil rubbing once a week;

use flannel cloths only on parts of your body most sensitive while exposed to the cold air.

Remedy for a Failing Memory.

Several correspondents have written for prescriptions adapted to restore the memory. We will answer all of them, in general terms, that failing of the powers of memory is a symptom of inactivity and decay in the thought-substance on the surface of the brain. The true remedy consists in a reinvigoration of the digestive organs by a temperate use of food. It will help, to sponge off the surface of the body in cold water every morning. Great care must be instituted about the appetite. Thousands eat themselves out of all their most retentive faculties. Dyspeptics are invariably disqualified for the retention of beautiful memories. They recollect distinctly what did *not* suit them at the last meal. If you want to put your brain in good condition, never load your stomach with anything. Neither smoke nor chew tobacco.

How to Overcome Morphia.

We have heard from many victims of Opium, Morphine, Snuff, &c. Some deny the existence of any power of reform in themselves. Others think they have the requisite Will-power, but plead ignorance of its application, as the principal reason why they do not apply it instantly. A suffering lady, under the dangerous mastership of Morphia, has written for information and remedy.

REMEDY.—Resolve to live a natural life or die in the effort. Be inspired with great power of Will. This is the beginning of redemption. Refuse to take another portion of the Master. Put it beneath your feet; plant your heel on the Serpent's head; arise, loved patient! in your spiritual might, and to Morphia say: "Get thee behind me, Satan!" Having passed the hour for taking the customary portion, your only next step is to reduce the quantity of food at least one-half, and do not increase the amount at any subsequent meal, for many days, although the pangs of hunger may smite you severely at times. If a "gnawing" is felt, take a little lemon juice on sugar, or drink a small cup of strong coffee without milk and food. Avoid fruit between meals. Green fruit, well cooked and not very sweet, or a tablespoonful of Indian meal, in a cup of water, every morning, will act as a cathartic and diuretic. Lastly, provide yourself with the homeopathic preparation of camphor, and, for a few days from the hour of resisting the evil, take a portion of it as substitute. Remember, you have the power of "Self-Healing" treasured up within your organism. Evoke it at once; believe on it; and the gods will cover you with their benedictions. [We have recently heard from the "suffering lady," just alluded to, assuring us that *she did conquer the habit.*]

Treatment for Nervous Debility.

In cases where the system is chronically afflicted with a nerve-debilitation, the symptoms varying with changes of the season, but never quite removed, we prescribe a complete and thorough change in the habits of life. For fashionable persons, in independent circumstances, the adoption of habits of working, eating, and sleeping, like farmers or mechanics, is advisable; and, *per contra*, those who live by the sweat of their brow, should be permitted to rest a few weeks from their labors, aided meanwhile by the abundance of the wealthy, until their worn and weary bodies can recover somewhat of natural tone and power. Let everybody work enough to deserve the food he eats and the raiment he wears; let no one be compelled to earn sufficient to support another in idleness; the sick, the aged, and the infantile, are naturally exempt from labor; but no Health can exist in a world where two-thirds are under-fed and over-worked, and where the residue are over-fed and habitually cloistered in mansions of luxury and indolence. If all should work some every day, "the world would be the better for it." Doctors and ministers could very soon be spared from such a world.

The True Elixir of Life.

A correspondent of the *Knickerbocker* gives the following testimony : "I am a WOMAN : have been through all the 'experiences;' hooping-cough, measles, dyspepsia, 'nerves,' 'blues,' *et cetera ;* have 'died daily ;' and, at last, come to life, health, and happiness, having found the 'Elixir of Life '—*Exercise and Fresh Air.* I began with the homœopathic dose of half a mile ; felt the thumb-screw torture at every joint ; persevered ; and now count ten miles a trifle, in all weather, and at all times ; never take cold, dyspepsia, 'nerves,' or the doctor's stuff ; and in consequence, never get cross."

Depression of the Vitality.

Many persons complain of a depression in all the vital forces. The patient is frail and failing from a general deterioration of the functional processes. Your true medicine is human magnetism, but circumstances may be against you ; therefore, as a *revivifying* tincture, we prescribe : One quart of Holland Gin ; put in it caraway, fennel, and mustard-seeds, of each, bruised, one ounce ; gentian root, mandrake, and thoroughwort, of each, bruised, half an ounce ; then put it all in a bottle or jug, and add half a pint of olive oil and a table-spoonful of finely-powdered cloves. Always shake before using. DOSE.—A teaspoonful or more after awhile, whenever your weariness is most difficult to bear.

Cause for Aphtha Chronica.

The patient, if a child, is *very* weak, and poor, and puny. He vomits much. His food seems to do him but little good. His flesh is soft and flabby, and the skin looks dead. He sweats nearly all the time when asleep.

The depleting cause of this disease is in the nervous department. The ganglionic forces are weak. There is, consequently, a preternatural *apathy*—a morbid suspension of the energy of the brain in relation to the digestive processes—and, perhaps, an *aphthous* condition of the whole alimentary canal. Physicians usually expect to find evidences of this disease upon the tongue and gums, inside of the lips, and on the palate, in the shape of little white ulcers. With these indications, the disorder is commonly termed "thrush." But *aphtha lactucium* is the best and most comprehensive word to convey the whole meaning. Old persons are sometimes thus affected just before physical death. It is often a final symptom.

REMEDY.—Make a little pudding, two or three times a week, of equal parts of rice, barley, wheat-flour, and hearts of mullen, all perfectly pounded and macerated together, and boil the whole compound until it is thoroughly done. This will be a curative food for the suffering. It may be given with milk, or in sweetened water. An onion poultice should be applied over the stomach *in the early part of each day.* Let it remain on about two hours. The onion is capable of invigorating the prostrated nerves of the digestive system. Let the patient sleep with the body, from the hips upward, considerably elevated. Never give it food more than four times in twenty-four hours

Derangement of the Liver.

The patient complains of great languor and low spirits; brick-dust deposit from kidneys; inclined to costiveness; sour stomach frequently; complexion yellow, with brown spots here and there on the face; no great suffering, but much depression and weariness.

CAUSE AND REMEDY.—The liver is the sugar-making organ in the human economy, and is the first of all the visceral confederates to report any deficiency or excess in this particular. The above symptoms prove that there is an excess of bile (or old, worn-out, broken-down blood globules or eggs,) and that there is too much sugar and alkali sent out from the liver. The condition of the kidneys and stomach, taken in connection with the pallor and languor, are evidences unmistakable. The true *remedy* consists in a cheerful spirit (a very difficult medicine to obtain under some circumstances,) and plenty of nutritious food, which must be selected wholly from the vegetable kingdom. For supper eat no stewed berries or fruit. Nothing sweet, nor sour, nor salt, but only a tumblerful of Indian meal porridge, with a little unbuttered bread. Retire very early; arise, if possible, before the sun; walk out, breathe, and *think* of the Summer Land. Cultivate the spirit of Hope. Truth is more lovely than persons; worship it; let them not enter the sanctuary of Thought. For breakfast drink one cup of black tea (*sans* milk or sugar,) and eat nothing. Let hunger assail the citadel. Do not yield. For dinner eat whatever, and as much as, your appetite demands. Never sleep after dinner. This course of treatment will effect a cure in about four weeks. Take a mild cathartic two or three times.

Torpid State of the Liver.

The symptoms are as follows: The eyes seem to swell in their sockets. Severe pain in the front and lower portions of the brain; giddiness. Extreme sickness ensues; vomiting of bile, and discharges from the bowels, until the lower system is depleted. Then you are prostrated, unable to labor, only able to sleep and groan. Food is loathsome; and your flesh is wasting slowly away. You have these turns from six to ten times a year. Quarts of medicine have flowed down your throat. Powders, pills, plasters, poultices, and other pestilences, have been used in your behalf. You have no faith in the Water Cure, because you have " tried it;" no faith in the Old School, for you have " paid it" all your dollars; and you, therefore, want—

THE REMEDY.—This is very simple, and was from the beginning—to wit: during the period of comparative freedom from the symptoms, let your friend *rub* and *pound* the parts most affected. Frictionize just over the liver; manipulate the stomach and bowels; chafe the skin upon your back; and lastly, squeeze and smite the legs and arms. After this process is completed—which should be in the early part of every day—let the same hands anoint your surfaces, and then wrap them in thick flannel. Previous to the manipulations, it is proper to wash and cleanse the skin of its perspirational accumulations. Rabbit's oil is the most penetrating for this purpose, but olive oil will answer, if mixed with a small quantity of oil of spearmint. In addition to this, take charcoal mixed in Holland gin. The ancients were not unscientific in the use of sweet-smelling oil upon the body.

Remedy for Persistent Biliousness.

Many want to know how to overcome a constant inactivity of the liver, biliousness, jaundice, headache, fever, chills, &c., which abound in many parts of this country. REMEDY.—When the system is clogged and surcharged with broken-down blood, wasted tissue, semi-oxygenated fluids (out of which corruption, all manner of jaundice, biliousness, headache, melancholy, fever, chills, &c., are brought forth,) the only remedy consists in a persistent and judicious course of cathartic treatment, such as rhubarb, and charcoal in gin or water: a tea-spoonful of each, every night for, perhaps, a week; then every other night, for some ten days or longer; lastly, once a week, as long as there remain any symptoms of torpid liver, or any disease whatever. If the operation of the medicine produces weakness, yield to it, and rest both day and night. Persons should use less or more of the rhubarb, according to their intestinal susceptibilities. Dietings, nursings, and milk and water treatment, in incorrigible cases of biliousness, are simply trifling with a formidable enemy.

How to cast out the Devil.

A mother, having a little faith in the Swedenborgian phase of Spiritualism, and considerable more faith in "good, old-fashioned New England Presbyterianism," is alarmed for one of her eldest children, who is a partial medium, thinking it possible that some "evil spirit" has taken possession. The symptoms are variable, but the following is given as a synopsis: Occasional flushes, or a circumscribed spot on one or both cheeks; the eyes become dull; the pupils dilate; an azure semi-circle runs along the lower eye-lid; the nose is irritated, swells, and sometimes bleeds; occasional headache; an unusual secretion of saliva; furred tongue; breath very foul, particularly in the morning; appetite sometimes voracious, with a gnawing sensation in the stomach, at others, entirely gone; occasional nausea and vomiting; violent pains throughout the abdomen; bowels irregular, at times costive; belly swollen and hard; urine turbid; respiration occasionally difficult, and accompanied by hiccough; uneasy and disturbed sleep, with grinding of the teeth; temper variable, but generally irritable.

CAUSES AND REMEDY.—Our decision in such a case (religious predilections being set aside,) is, that the patient has not been well educated in matters of eating and drinking; has from childhood used too much sugar, ate bread made with saleratus, too many buckwheat pancakes; and that, as a sad consequence, the patient is a "medium" for evil and diabolical "spirits" in the horrid form of "*Worms.*" We cheerfully give directions for exorcism of the invaders. In order to successfully "cast the devil out" of your child, first reform the diet, abolishing bad bread and all sweet articles of consumption; secondly, give the sufferer a cup of tea made of a little of each—sage and sweet fern—every forenoon; or rhubarb and charcoal once or twice a week. May the "devil" soon depart both from your creed and family!

Dyspepsia and Despair.

This is a disease of the mesenteric glands of the ganglionic nerves in the digestive system, affecting the intellectual and better faculties of mind. Depression and mental dyspepsia are natural symptoms under these conditions.

REMEDY.—Evoke your whole manhood! Never say "die." Take a tea-spoonful of *Charcoal*, in hot water, twice or thrice a week. Take a lemon, cut off the end, fill it with white sugar, and then slowly squeeze the contents into your mouth. This, and nothing else, is your last meal. Your nights will soon become periods of rest and invigoration. Rise early. Sleep before breakfast, if possible. Your morning meal must not be watery; nor your dinner; neither should you *ever* taste a particle of fruit or berries between meals. Thoroughly oil your whole person with sweet oil, perfumed as you like it, once per week. When thirsty, use lemon and sugar; not anything stimulating, nor cold water. Believe in Nature's remedies—they will not fail you.

My Dyspepsia and my God.

Alas! madam (said a plain-minded deacon,) I have seen too many souls go to perdition by what you call "Health Reform." No sooner has a person quit coffee, than he disbelieves in Infant Baptism; with tea goes his reverence for the Eucharist; let him leave off eating pork, and he will discard the doctrine of Vicarious Suffering; let him take no more medicine, and he stands in danger of the heresy of Universal Salvation; and by the time he is a finished vegetarian, he will deny the doctrine of Plenary Inspiration, and drift straight into the quicksands of Infidelity. No, madam, give me, rather, my dyspepsia and my God!

Looseness of the Bowels.

Many persons are afflicted with an almost continued looseness of the bowels; sometimes troubled after meals, or whenever the least excited.

REMEDY.—The Pneumogastric Cure will perform wonders in all such cases; for the stomach and bowels throughout are wholly under the scepter of the sympathetic and pneumogastric nerves, through which the Will-power works like a lever over its fulcrum. Lest, however, the patient is deficient in volitional ability to "heal" himself, we prescribe for all ordinary forms of summer complaints, relaxation of the bowels, diarrhœa, &c., simply and only this: *Chew a few cloves every day*—better between breakfast and dinner. Avoid drinking largely between meals. This remedy will cure even chronic cases of looseness.

Heat and Pain in the Bowels.

There is ofttimes a subdued inflammatory disease of the muscular texture of the intestines—almost *enteritis;* also a tardiness of action in the mesenteric glands, causing weakness and distension to result from much eating; defective chylification at times; an almost neuralgia of the sympathetic ganglia throughout the bowels.

REMEDY.—Beware of everything which tends to *entaticus*, (physical irritation.) In your programme of eatable articles, the following must be omitted: "Old smoked salt meat, salted fish, veal, geese, ducks, the liver, heart, lungs, or tripe of animals. Rancid butter, old strong cheese, lard, fat pork, turtles, terrapins, oysters, raw or cooked, hard-boiled eggs, omelets." Adopt this plan, but eat regularly of whatever else suits your taste. Envelop your bowels in oiled silk every night, or rub with sweet oil, and wear an apron of fine fur on the surface of your abdomen all day. Let every person afflicted with neuralgic pains in the stomach and bowels adopt the fur-remedy at once. It is equally good for coldness of the intestines, tardy digestion, and habitual flatulency. But manipulations and the Will are indispensable in all cases.

Cure for Sick Headache.

For years some poor soul is severely afflicted with a nervous sick headache, which gradually wastes his body away, and brings his existence to an end. He has tried every remedy within his reach, without relief. REMEDY.—Such a case is not hopeless. We almost know that it can be tuned up and made healthy—on condition, of course, that the patient will strictly follow the prescription hereby imparted. First. He must particularly scrape the tongue every night and morning. Remove all *debris* of the salivary glands, and wash away from the mouth all miasmatic epithelium, so that the tongue shall be clean and capable of accomplishing the first offices of digestion. Second. He should drink nothing stimulating of the alcoholic character, must not eat as much meat as he may sometimes crave, and, particularly, he should fast at least twelve hours previous to the usual period of attack. Third. The sovereign remedy may be confidently administered —namely: Drink a glass of *very sour buttermilk* instead of supper, and repeat the dose a few minutes before breakfast. The milk should be used at night only a few times previous to the usual sickness. The object is to remove the febrile action throughout the body, which is the cause of the progressive emaciation, and nothing can do this but a remedy that shall supply *lactic acid* to the blood, and at the same time overcome its extreme alkalinity. In most cases this remedy will prove a cure for sick headache, more especially when the stomach is sore and the kidneys weak.

An Inveterate Dyspepsia.

Pain soon after eating breakfast, bloating immediately after dinner, and "a horrid nightmare almost every night," or a dreadful headache about every third day. REMEDY.—Eat precisely that which you know gives you the least suffering, and as much of it as you want; but omit your supper, without deviation, until completely cured. Instead of tea or supper, eat a lemon, only a little modified with white sugar. Before breakfast, drink a tea-cupful of sour buttermilk. We think that this course will cure you, because your dyspepsia is wholly owing to a deficiency of *lactic acid* in the system.

Remedy for Bilious Vomiting.

Persons afflicted with bilious vomiting, and sometimes diarrhœa, should eat *nothing* after dinner. Let this be a rule day after day for months. Hunger will be a good medicine. First thing on rising, mornings, drink a tumbler of cold, weak, black cherry bark tea, which may be prepared during the preceding evening. A little powder of camphor, or of salt, laid on the tongue after the vomiting commences, will stop it, and prevent diarrhœa.

Swelling of the Abdomen.

A Boston lady describes her case as follows: "For a few months past I have been troubled with a swelling of the abdomen, and hardening at the same time. There is no pain at all accompanying this swelling, but it alarms me. It occurs nearly every day—sometimes, indeed, more than once a day; is *never* larger than usual in the morning before rising, but as soon as I bathe myself, it swells and hardens almost before I can get dry."

Cause and Remedy.—This is the second instance of a peculiar affection which has come before us in the department of human disease. The anterior surface of the intestines is covered with adipose matter, called "omentum," which, in persons of fleshy proclivities, is abundantly deposited between strata of the serous membrane. The mesenteric glands, through which the lacteal vessels pass to the thoracic duct, are somewhat inflamed. From thence a *gas* is rapidly diffused between the intestines and the membrane that contains the omentum. The sudden swelling and hardening are natural concomitants of this intestinal transaction. In one case, where this condition was permitted to exist for over two years, the patient died with a tumor (fleshy and fatty in composition,) in the abdomen. Another, with identical symptoms, recovered by abandoning all drinks at meals, and taking, every forenoon, a very warm sitz bath for fifteen minutes. For every such bath make a strong decoction of hemlock boughs. It is essential for the patient to perspire freely in the bath.

Water in the Stomach.

In most cases, we observe that the accumulation of *water* in the side, stomach, bowels, or about the heart, is caused, in the first place, by the derangement of the liver, and, in the second place, by generation of *gas* in the lower stomach and small intestines. Remedy.—Drink nothing between meals, and no more than a tumbler of fluid while eating. Twice a week, take a table-spoonful of powdered charcoal in a wine-glass of warm water, about one hour before dinner. It may be necessary to rest or sleep a few minutes while under the immediate influence of this medicine. For "Prairie Itch" there is nothing better than the tea of yellow jessamine; also wash the body with strong decoctions of lavender and hemlock bark.

Cure for Habitual Costiveness.

A constant constipation may be cured without recourse to artificial means. Graham bread and plenty of apple-sauce for breakfast; no meat, no hot cakes oftener than twice a week; no coffee at any time; and very little fluid of any kind. This method is adapted to all persons whose occupation keeps them within doors. For an immediate relief, take a table-spoonful of Indian meal or Graham flour, in a tumbler of water, before breakfast, quite early in the morning. Perhaps several doses will be necessary.

A Bilious Medicine.

The system of every bilious patient is radically impaired It must be built up by the most persistent effort of Will. In addition, we prescribe a bitter medicine, as follows: Mandrake root, pulverized, one drachm; orange-peel and cloves, of each one table-spoonful, well pounded. Put in one pint of good brandy, one pint of water, and one pound of brown sugar. Stand one week; then add another pint of water. DOSE.—Commence with a tea-spoonful before meals.

Evils of Eating for Amusement.

A Newburyport (Mass.) paper says: "A young man residing not a thousand miles from Beck-street, being disappointed in going to the Bluff, last Monday, consoled himself by consuming the following refreshments, in addition to three hearty meals: Five sheets of gingerbread, three glasses of small beer, five glasses of nectar, three large pickles, twenty cocoanut cakes, six ounces chocolate cream drops, ten cigars, seven large apples, half pint of peanuts, four cents worth of old cheese, one stick of candy, one pint of new milk, four glasses of ice water, and an *emetic*, which was ordered by a physician, to save his life for further duties."

Tone of the Stomach Destroyed.

When your stomach is inclined to burn, and to refuse every ordinary article of food and drink, the true remedy is hand magnetism, applied to the spine and over the stomach. Drink the mildest tea of roast onions occasionally. Rye bread, well toasted, is better than wheat for a weak digestion. Swallow uncooked Indian meal, or chew wheat berries

Your diet should henceforth be more nutritious. Not either "fish, flesh, or fowl," but the grains that grow in the sunlight. Make a pudding of equal parts of Barley, Wheat (cracked,) and Corn. Eat this as the principal article for your dinner. Abandon desserts of every description, and take a light breakfast.

When the stomach continues weak, with weariness in the fore part of the day, but a better state of feeling as evening approaches, we recommend the patient to eat nothing hearty till dinner. Sleep, an hour in the forenoon, is particularly useful.

A New Test for Diabetes.

Professor Paine, in his *Journal of Eclectic Medicine*, says: Drop one or two drops of the urine upon a slip of clean tinned iron, hold it over a fluid lamp, evaporate the fluid, and continue the heat. If there is sugar in the urine, a rich, reddish-brown color will appear on the place from which the urine is evaporated.

Remedy for Urinary Weakness.

Some children, as well as adults, have an inveterate habit of wetting their beds at night. It may be proper to denominate this disease *Diabetes insipitus*. It is caused by a superabundance of serum in the blood, which contains a too large proportion of saccharine matter, unassimilated. The urine at first is clear and sweetish, but very soon gives off vapors peculiar to the general condition of the system.

REMEDY.—Hygienic means are always in order, and essential to a cure. That is, the unfortunate victim should eat or drink nothing sweet or sweetish; nor may the stomach and bowels be fed with starchy food, such as potatoes and fresh bread. A morbid state of the blood is the cause of the weakness in some children; but in nearly all cases the primary cause is a too frequent use of milk and sweet diets, puddings, etc. No fluid should be taken with meals, or at any other time, unless the thirst is intense, in which case use strong *lemonade* without sweetening, or a tea-cupful of water, medicated with from one to three drops of diluted sulphuric acid. Twice a week the body of the patient should be thoroughly anointed with sweet oil, dissolved with a little of each—spirits turpentine and alcohol. Always bathe and manipulate from head to feet, except when the patient has some local inflammation, or special pain. ☞ The disuse of all starchy and sweet foods, the use of the WILL as both a policeman and a physician, and the acid drinks recommended, is a good treatment for the great majority of cases. Severe and long-standing sufferers with incipient *Diabetes*, may aid their cure by putting one ounce of cascarilla bark, one drachm each of cloves and cinnamon, pulverized, and four ounces of lemon-peel, in one pint of best port wine. To tincture one week. DOSE.—A tea-spoonful, with a Graham cracker, for supper.

How to Stimulate the Kidneys.

A bilious state of the system is sometimes attended with an obstruction of the flow of urine, which tends rapidly to aggravate all the symptoms, and to inaugurate the disease called *jaundice,* which, in turn, is not unlikely to affect the hepatic functions, and ultimately the lungs.

Remedy.—There are many articles in the vegetable kingdom calculated to stimulate the kidneys. Balsams, turpentine, dulcamera, digitalis, juniper berries, spirits of nitric ether, &c.; but the most popular with a few practitioners over the Atlantic, is a remedy called Stork's-bill (*erodium cicutarium,*) which grows in several parts of this country. Dr. Bryerley, of Cheshire, England, gives, in the *Medical Times and Gazette,* the following directions for its use: " The mode of preparation is, to infuse an ounce of the dried plant (every part of it,) in three pints of water, stewing it in an oven until two pints remain. The dose for an adult is four or five fluid ounces three times a day; probably more may be needed in some cases. The Stork's-bill is indigenous in England, where it grows abundantly on sand-hills near the coast, but it has been introduced into this country, and is to be found on the shores of Oneida Lake, in the State of New York."

Medicine for a Weak Stomach.

Sweet fern, if pulverized and chewed about an hour before dinner, would prove higly beneficial to a weak stomach. Swallow the juice only; not any of the herb. Sweet flag is not good for nervous characters. A sweet disposition, used abundantly between sleeping hours, is a royal remedy.

The roots and berries of (*Aralia racemosa*) spikenard are sometimes efficacious in dyspeptic disorders. Botanic and Eclectic physicians attribute *Stomachic* and mild *Balsamic* or stimulating properties to this common herb. The alkaloid extract is occasionally used. We have found it useful only when combined with two or three other medicinal herbs of more power.

There are two ways of preventing and of curing dyspeptic conditions—the manly and the mean; the manly by going to the table twice a day, and nobly curbing the beastly appetite, saying: " I will eat this and so much, and no more by a single atom." The mean or ignoble, by having " this and so much, and not an atom more," sent to a private table; the " this and so much," the quality and quantity, having been determined by the observed instincts and needs of the system; each man being a rule for himself, under the guidance of a wise physician, or of an unerring and competent judgment of his own.

Remedial use of Sugar.

A little reflection will inform you when, and in what cases, the use of sugar or syrups is consistent with health. Sugar may be advantageously used only whenever patients are not diseased in their nutritive functions. Digestion is sometimes injured by the too free use of sweets. For this reason we do not often prescribe " syrups " for our patients; as, in this country, the majority of diseased persons are dyspeptics, or greatly debilitated in their organs of nutrition. In the stomachs of such patients sweets become *acids*, and thus many unpleasant symptoms are developed.

Cure for a Sour Stomach.

The causes are various. The remedy is simple : Use a few drops of pure lemon acid just before eating. But your food should be plain; no sugar, or sweets, or pastries; eat plenty of rye bread, and sometimes chew chamomile flowers.

Honey and other Sweets.

The nectar of flowers, gathered by bees, is a watery solution of cane sugar. In the process of this transformation, the cane sugar is decomposed into three different kinds, which constitute honey. The heat which the bees maintain in the hive causes this change. Weak acids, as well as heat and moisture, can effect a similar conversion of cane sugar. Your digestive organs cannot, in this northern climate, dispose of large quantities of sugar. Health demands very little of either sweet or sour. Children do not need more sweet than is contained in the milk and fruit they everywhere naturally consume.

Sweating of the Extremities.

This is generally produced by an inactivity of the absorbent system. The mesenteric glands are diseased, causing symptoms not unlike indigestion and torpid liver, but the effect on the nerves is debilitating and harassing. REMEDY.—Horse-radish and Turkey rhubarb, of each one ounce, infused in one quart of good brandy for one week. Then add one pound of brown sugar, one tea-spoonful of powdered cloves, as much red pepper, and mix by shaking frequently. DOSE.—One tea-spoonful, with an agreeable quantity of water, whenever the sweating of hands and feet is most profuse and troublesome. You should never take more than three doses of this preparation in twenty-four hours.

Worms in Children.

Some children have a predisposition to this distressing and disgusting form of disease. The long round worm, or *lumbricoid ascaris*, is very common. The tapeworm is rare, but dangerous. The usual *symptoms* are griping pains in and about the abdomen; variable and voracious appetite; fetid breath and occasional nausea; wasting away of the body; disturbed by troublesome dreams; dizziness; bloating of the stomach; pain and itching at the navel; grinding of teeth while sleeping;

picking and irritation of the nose; and sometimes slight convulsions.

THE REMEDY.—Stop all sugar and sweets of every description; no pies, puddings, cakes, or preserves; bathe the stomach and bowels in diluted oil of cloves. Mix a little of finely-powdered cloves with whatever the child eats for supper. Weak clove tea is excellent. But the quickest and least dangerous remedy is an injection of a tea-spoonful of linseed-oil in a sufficient quantity of warm water. This may be used twice or thrice a week. Constipation must not exist. With costiveness, a cure of the depraved condition will be impossible.

Adult intestines are not exempt from such verminous visitations. We shudder when we recall what we have witnessed in human bodies. Vermifuges are numerous. For an adult, three grains of pulverized Indian Hemp (*Apocynum Cannabinum*,) twice a day. Male fern tea, for children, or tansy and wormwood steeped together, and drank every other day. We would recommend injections of a weak decoction of cabbage-leaves.

The Prepared Female Organism.

QUESTION.—"MR. DAVIS: In your answer to a correspondent's question, 'Why are not men and animals produced now as they were at first?' you say: 'We do observe a time when the highest animals started through the reproductive organism of the prepared female,' &c. Now, the question I wish answered, is this: What do you mean by the *prepared female?*"

ANSWER.—Readers of the volume of Nature observe that each chapter of material development is marked by deep-reaching changes in the fluids and solids of the globe—vast crises and earth-wide revolutions—accompanied by the retirement or extinction of one set of physical conditions, and followed by the inauguration of new and superior circumstances in the material constitution of things. These changes, or crises, or revolutions, or whatever you wish to term these *transition* points and passages in the globe, are far more perfect, and, therefore, less conspicuous and less remarkable, in the world of organized animation. By Clairvoyance we anticipate the results of scientific discovery, which will be this doctrine of the origin of the human species: That mankind came not from the progressive transformation of the physical organisms of the superior animals or *Troglodytes*, but by and through the advanced *reproductive organisms* of females of the ante-human types, which had, in this particular respect, arrived at a fruit-bearing *crisis* or *change* in regard to procreation, whereby a higher type (the first human organizations,) entered upon existence. The particular philosophy of all this will be explained in our little volume on "The Reproductive Organism."

The Presence of Milk.

QUESTION.—Dr. J. P. C. ...., IOWA, relates a remarkable case, and asks whether the presence of milk in the breast is not an infallible sign of an advanced stage of gestation and pregnancy?

ANSWER.—No; the secretion and accumulation of milk in the breast is not an infallible sign, because there are many well-established exceptions. And yet, as a great general principle, the fact of pregnancy is invariably accompanied by this kind of evidence. Man's organization has been known to secrete and convey milk to one of the breasts, and thus to support life in a very young child. Captain Franklin, in his interesting narrative of his journey to the shores of the Polar Sea, relates the case of a young man, of the Chipewyan tribe, who, having lost his wife three days after giving birth to a son, actually nursed the child from his left breast, and long subsequently it was ascertained that he had discharged the duties of wet nurse toward several little members of his son's family. He performed these offices, so unbecoming to his sex, from gratitude to the "Great Master of Life" for permitting him to save his motherless infant son.

Cure for Inferior Desires.

QUESTION.—"Is there anything in the Materia Medica that will prevent sensual thoughts from rising in the mind?"

ANSWER.—This disease, in the majority of cases, is hereditary. Being wholly ignorant of the direful consequences, thousands encourage depraved imaginings, until character is malformed and virtue is swept from the sphere of private life. No system of medicine is master of the human system. All pretensions of physicians to cure reproductive diseases, *are impositions.* Otherwise, the word would be: "Go heedlessly forth —sin every day—you shall not suffer."

With great success, however, we have prescribed, for the truly repentant, two grains of *African Capsicum* (cayenne-pepper,) made into a pill, and swallowed the last thing on retiring every night, and continued for several weeks. For the young, this prescription is particularly effective and useful. Sufficient employment, mild diets, cold water bathings, and contemplations of Nature, are all remedial. The sovereign remedy is a true conjugal life, wherein love—not passion—rules the soul with an unerring government.

Stop the use of cane sugar in every form. Take no drinks, cakes, or puddings, which are sweetened with sugar or treacle. Grape or rose saccharine is good. Hence honey is not injurious in this disease. Another way to restore the seminal vessels to their original strength, is, to chew plentifully of the leaves and flowers of chamomile during the forenoon, only swallowing the wine and saliva, which, by the process of mastication, you extract from the quid.

Weakness of Bladder and Kidneys.

Bathe your back and throat in cold water every morning From breakfast to dinner chew prickly-ash berries, or bayberry bark, either of which will cure a weakness of the bladder and kidneys. A throat affection is sympathetic.

If you have often a weakness through the small of the back, which prevents you from doing anything that requires strength, there is an incipient disease in the kidneys—a wasting process inaugurated in the cellular tissues—which must be stopped at once. Sweets must not be taken; no pastries; nor hot drinks of any description. Get Thimble Weed (*Rudbeckia lacinata,*) and chew it plentifully every afternoon, swallowing only the juice and balsamic properties. Frequently bathe the loins in cold water. Every night wear on your back a cold compress (a bandage wrung out in cold water,) and well protected with woolen cloth of much thickness. Disuse all *very* salt food; no fresh-baked bread; plenty of fruit, if well-cooked but not for supper.

Cure for Fluor Albus.

It is not necessary here to enumerate the primitive causes that ultimate in Leucorrhœa. Every intelligent young lady is sufficiently enlightened on physiology to know that the increasing discharge of a white secretion from the internal membranes is certain to result unhappily in the after years, which should be full of joy and beauteous health.

REMEDY.—Use a syringe every morning of white of an egg, thoroughly amalgamated with a table-spoonful of liquid honey. Pour this mixture into sufficient blood-warm water to make over half a pint, for injection. At night take a one-grain pill of the extract of chamomile. After one month, if your symptoms are not removed, substitute for the above injection one pint of blood-warm water, containing from three to six drops of diluted sulphuric acid. In a few weeks get your druggist to make you an hundred pills, of two grains each, composed of equal parts of extracts of chamomile and dandelion. Take two every other night. Obey all the laws of Nature, and so realize a full unity with the Spirit of your Father God.

Cure for Vaginitis, or Irritation.

All women, young or advanced, may treat the symptoms of this affection, and cure it. The term "Vaginitis" is applied by pathologists to that very prevalent *irritation* and painful inflammation of the vagina, caused by some form of the disease known as leucorrhœa. The *Bulletin de Therapeutie* publishes the most reasonable prescription for this painful affection, consisting of eighty parts of glycerine and twenty of tannic acid.

When the vaginitis first appears, the inflammatory symptoms should be calmed by appropriate regimen, baths, and frequent emollient injections. When the first stage of the inflammation has passed away, and the careful introduction of the speculum has become possible, abundant injections of water are to be thrown in, so as to remove all the muco-pus which lines the walls of the vagina, and these are then dried by a plug of charpie, placed at the end of a long forceps. Then three plugs of wadding, well soaked in glycerine and tannic acid, are to be introduced. Next day, after a bath, the plugs are removed, new injections made, and the dressing repeated. M. Demarquay has never had to have recourse to more than four or five such dressings. After discontinuing them, astringent injections, consisting of infusion of walnut-leaves, in which one drachm of alum to the quart has been dissolved, are employed two or three times a day for a week or ten days.

Cure for Painful Menstruation.

For painful menstruation and low circulation, we prescribe magnetic manipulations. If you cannot avail yourself of this treatment, then, as a substitute, wear *a fur garment* below your waist, enveloping the entire abdomen and the hips, which should be put on and worn constantly about three days before the flow commences, and taken off two or three days subsequent to the cessation. If your feet and body are painfully cold, after retiring for the night, your best remedy is very long fur stockings, and mittens of the same, with neatly-fitting wristlets. Wear these only at night. Fur soles within your shoes during the day, or moccasins. Lamb-skins, with wool next to your body, will answer in place of certain kinds of fur. The fur should always be worn with the flesh-side inward.

Whenever menstrual pains are severe, or your body is more than usually weak, put across the small of your back, over the kidneys, and reaching to the loins, a girdle of flax-seed poultice, well sprinkled with pulverized camphor.

Displaced Uterus.

The throat diseases, and many other symptoms which afflict you, are caused by the derangements of the reproductive organisms. REMEDY.—Let some lady physician adjust the organ to its appropriate place about two days before menstruation. Do nothing that will cause a recurrence of the misplacement. Take the Spring Beverage as prescribed. Use your Will whenever you walk in doors or out.

Cure for Spina Bifida.

Physicians have classified this disease into dropsy of the spine (*Hydrorachis*,) and an incomplete state of the vertebra, giving rise to lumps and tumors under the skin. As to the treatment, we know of nothing comparable to the electro-magnetic battery in connection with hydropathic hygiene. If taken in time, and skillfully manipulated daily, either form of Spina Bifida would get well—providing, of course, the patient has a constitution strong enough to furnish the healing energies.

The Source of Marrow.

The marrow (scientifically called *medulla*,) is digested through diverse cellular membranes, and then, by a secretory process of the small arteries, it is deposited in the appropriate cavities of the cylindrical bones. It contains very little of the primal constituents of the organization, but, as a residuary and conservative substance in the hollow of the structures, a healthy marrow is indispensable.

Treatment for Lumbago.

This is a painful rheumatic affection of the muscles about the loins. It sometimes begins by stabbing the patient severely for a few hours, and then leaves the victim with a lameness of the most distressing character in the hips and loins. REMEDY.— Take care to keep the skin of the parts perfectly clean by means of soap and warm water. Wear a lamb-skin across the hips every day; sometimes, when the suffering is intense, keep it bandaged on all night. Let all sufferers beware of sensual indulgences.

Nervous Trembling at the Stomach.

It will be necessary for the patient to receive magnetic treatment, half an hour every day, from the hand of some congenial person, either man or woman. You should assume an easy position, close your eyes, and remain passive in spirit, while the operator's hand is laid upon the pit of the stomach, occasionally making passes from the stomach upward to the throat and around the neck, terminating under the back hair A powerful operator need not often move his hand during thirty minutes. We regard still-magnetization as far more efficacious when applied to the pit of the stomach. The left hand should be first applied.

Spiritual Afflictions and Discordant Spirits.

Perhaps your nervous system is exposed to the magnetism of discordant cogitations. They may invade your chambers of thought—

> "Like the pitchy cloud
> Of locusts, warping on the eastern wind,
> That o'er the realm of impious Pharaoh hung
> Like night, and darkened all the land of Nile."

And yet you may assert yourself as Empress: and you may conquer your invisible enemies. First. Take nothing from the castor with your food, except salt in small quantities; abolish meat, except fish and eggs, for twelve weeks; bathe your person in tepid water, or take a hot water bath, twice a week, removing all the *debris* that will roll off by rubbing; lastly, read an hour or two every day in some work on exalting subjects, and discipline your mind to contemplate spiritual themes, and your heart to pray (desire,) for the kingdom of Heaven on earth.

Magnetic Treatment for Intoxication.

A case may be beyond the control of medicine, but not of magnetism. The spirit of alcohol may have taken possession of the soul. It is like an enchanter's charm, dispossessing the Will of its mastery, inclining the patient to do that which is most repugnant to his higher convictions. Drunkenness is a *spiritual* disease—not a habit which may be broken at will—and the true remedy therefor is magnetism. Thus. Let some efficient operator persist until he gets the subject under mental control, which is indicated by his obedience to the mentally expressed wishes of the operator; then let the latter impress upon his brain a feeling of intense hatred of alcohol, at the same time inspiring his Will to resist the least approach of the enemy in any form whatever. This practice will be attended with great success, if the operator himself be a temperance man, and a lover of good for its own sake.

Beware of the Tyrant, Disease.

A young lady writes a volume in the following paragraph: "Brother Davis: The world is before me, and, in the buoyancy of youth, I build ethereal castles, that dwindle into nothingness before my more enlightened reason. Hopes and fears, joy and sorrow, are the computations of my every-day life; and I, therefore, long for more congenial conditions in the Angel World. My physical system is not in harmony, caused by obstructions that resist the influence of the usual remedies. No one can possibly prize the value of health who has never been sick. If I could once more regain my health, I should be as happy as the wild fawn that bounds o'er the western prairies."

Answer.—The most unnatural slaveholder is Disease. No tyrant was ever half so full of fierce mischief. His hands are red with human blood. He lives in our dwellings—nay, in our very bosoms, nestling in the warmest recesses of our hearts—destroying health and beauty before, and ofttimes *in*, our very eyes! How can we longer bear the presence of this sleepless wretch? If humanity should weep a flood of hot tears, for forty days and forty nights, the deluge of sorrow would not drown this serpent, this rampant monster, this great foe of all men, called "Disease." Drugs, doctors, ministers, cannot kill him; he is sovereign of them all, and of all the world besides. And yet there is one certain way to conquer him, to wit: *Obey the laws of Nature, and thus intrench yourself in Health.*

Dysentery, or Bloody Discharge.

All diseases, as we have many times told you, arise from disturbances in the magnetic and electrical dynamics (or forces) which pervade and regulate the corporeal system. Of this disturbance a *change* in bodily temperature is the first symptom or evidence. If the disturbance be a surface one—such as closing the pores of the skin, and taking a cold in the joints, and the like—the symptom is an inward fever, being an extra-heated condition in the vital parts. This condition increases the pulse, coats the tongue, stagnates the digestive functions, and reduces the desire for food, but increases the desire to drink. On the other hand, if the disturbance of the equilibrium commenced in the vital parts, such as oppression on the lungs from a cold, overloaded state of the stomach and liver, costiveness, and the like—then the symptoms are a surface feverishness, headache, stretchiness, sleepiness, accompanied with a very general depression and loss of bodily strength. In the summer season, as has been always observed in all hot countries, in camps, and over-crowded places, the most frequent effect of magnetical disturbances is exhibited in *Dysentery*, or bloody flux, and diarrhœa.

Symptoms.—Frequent griping pains in the intestines, with loss of appetite, sometimes nausea, chilliness and hot flashes, lassitude, and a little headache. The discharges are chiefly mucous at first, mixed with blood, but the passages are attended with griping pains, and, in some persons, a falling of the rectum, and a bearing down, as in bad cases of piles. There is a typhoidal state of the system—a wearisome fever, accompanied with a very exhaustive weakness, and, perhaps, some delirium. These are the characteristic symptoms of dysentery. We will now indicate the best home treatment, adapted to everybody and to every place, which will leave the patient free both of medicine and disease.

TREATMENT AND CURE.—The usual. but mistaken, course, is to give the patient a physic. The old (and yet popular,) school of doctors still resort to gentle aperients—rhubarb, magnesia, castor-oil, mercury, and diffusable stimulants. Many of the old line physicians will yet bleed for dysentery, if the patient be plethoric and corpulent, or inclined to apoplexy. Sudorifics, emetics, diaphoretics, and even blisters, for counter-irritation, in connection with laudanum and Dover's powders, are still employed by allopathists in the treatment of this very simple disease.

In order to cure the acute form of dysentery, without leaving bad effects in the system, *stop all food of every kind.* The greatest danger lies in eating. It is a mistake that you must eat to keep up your strength. You should remember that, in dysentery, several of *the finest vessels of the intestines are bleeding.* The entire mucous coating is overloaded with slimy perspirations and with negative exudations from the blood; and eating, therefore, is the most dangerous and most unnatural act, because perfect digestion is impossible. If it be necessary to give the patient anything "to keep up strength," let it be something mucilaginous, like the tea of flax-seed, gum-arabic water, slippery-elm tea, or the broth of mutton. Occasionally use a gargle of salted water, in order to allay thirst and dryness in the mouth. So much for food during hours, or, perchance, for days, or while the disease is upon you.

Next, as to bodily treatment: Take a warm Sitz-bath two or three times a day. Remain in the water from fifteen to twenty minutes. After drying the skin, which should be done briskly with the hand of some magnetist, if possible, cover the bowels with cotton, which should be well sprinkled with the pounded gum of camphor. Keep the bowels and stomach warm and comfortable by external applications—so also the feet and legs, by enveloping them in several folds of flannel. Be quiet, walk about but little, and always sit in an inclined position. If the feverishness is considerable, the patient should be washed all over in tepid water, should be rapidly dried with the magnetic hand of friendship, and then placed between fresh sheets. This treatment, with little or no homeopathic medicine, will cure the severest case of acute dysentery, in three or four days. We would urge one other thing, viz.: The use of small water injections about three times a day, or immediately after there has been a considerable discharge of mucous and blood, accompanied with straining and griping pains in the bowels. Retain the injection until Nature insists upon another discharge.

Treatment for Chronic Dysentery.

This disease may be controlled and cured by simple methods The first, as in acute attacks, is to stop eating. Make a tea of flax-seed, with a strong infusion of cloves; drink a swallow every fifteen or twenty minutes. Strong coffee, without milk and sugar, may be taken, with a little roast potato and bread, for dinner. Warm water, or slippery-elm injections, are not to be omitted in chronic dysentery. Bilious persons, having the obstinate form of this disease, will find great virtue in the following: Turkey rhubarb and willow charcoal, of each (pulverized,) one table-spoonful; of saleratus, a piece as large as a hazle-nut; put these in a tumblerful of water; let it stand covered up twelve hours, when, after thoroughly stirring it, the liquid will be ready for use. Dose.—A tea-spoonful of the liquid about once in every four hours during the day.

(☞ Never wake up at night to take anything in the shape of medicine.) We cannot too strongly urge the value of hand-magnetism in restoring the balance of health to the system. And, furthermore, we would once more impress you to remember that, especially in all stomach and bowel disturbances, the WILL is a powerful physician. Do not fail to avail yourself of his skill and benefactions. Always sleep with your mouth closed, so that the air, by passing through the warm nostrils into your lungs, may be purer and more magnetic, and, therefore, more energizing to the nervous system.

Diseases of the Heart—Hypertrophy.

A father says: "My daughter is seventeen years old. At times, suddenly, and without any warning, her blood seems to stagnate; she then falls to the floor, and we have to rub around the heart in order to bring her back to consciousness. After she comes out of this situation, she remembers nothing about it. Her body and face swell, and it is some hours before she can again pin her clothes together."

DIAGNOSIS.—The left cavities of the heart—particularly the walls of the left ventricles—are considerably thickened; and there seems to be a diminution of the capacity of the cavities, and not an enlargement, as is most commonly the case in this country. And yet the organ is not absolutely diseased; it is merely arrested, and spasmodically contracted now and then, while in the performance of its functions. If long continued, the result would likely be *hypertrophy* of the left auricle near where the pulmonary artery and the aorta overlap it, but this extremity need not be feared, because it may be avoided. The above symptoms may be classified under a general term—*Cardiac Syncope*—which means to strike down, or produce swooning, by disturbances proceeding from the organic action of the heart. It is possible that, occasionally, the membranes of the lungs and heart would enlarge with an excess of watery blood, but this is unimportant in administering the correct prescriptions. The primary cause of the whole disturbance lurks in the pneumogastric and sympathetic nervous systems. *The brain*, not the heart, *circulates the blood*. The heart is merely a *moderator*—an organ to regulate and graduate the operation. Hence the treatment should be addressed to *the seat* of those forces by which the blood is made to flow through the organism.

REMEDY.—Numerous remedies hasten to present themselves from the fields, water, and air; but we must give the *formula* that is best, and most likely to be followed: Mandrake-root (*Podophyllum peltatum*,) Fox-glove (*Digitalis*,) and Hardhack-leaves (*Spira tomentosa*,) of each half an ounce. Let these articles macerate (soften,) over night, in one pint of rain-water. Then steep over a slow fire about two hours, and strain off the liquid. When entirely cold, add one pint best brandy and half pound of brown sugar. DOSE.—This simple medicine is very powerful, and must be used with care, as an overdose of it would bring on the symptoms which it is here given to cure. The largest dose is a tea-spoon two-thirds full; it is best to begin with only ten drops night and morning. When the patient is attacked with insensibility, convulsions, and coldness of extremities, and swelling about the waist, do not fail to make hot applications to the skin over the heart and about the waist. It is well to rub and manipulate the chest and spine a quarter of an hour each night. Feet must be kept warm and dry, and rubbed with both hands, and then wrapped in flannel every night. Drink little fluid; eat no uncooked apples; one roasted potato, with brown bread and butter, should do for breakfast. Practice deep-breathing about ten o'clock every morning; likewise a few minutes after retiring for the night. The Will-power should be frequently exerted on the muscles of the stomach and bowels.

Palpitation of the Heart, and Wasting.

The great destroyer of the tissues, called Oxygen, works upon food, when the system is in a healthy condition; but, in diseased conditions, Oxygen co-operates with the alkalies, and thus consumes the tissues. This warfare is going rapidly forward in all inflammatory and febrile disorders, resulting in emaciation and in the perceptible *wasting away of the bodily substances*. But when the digestion is normal and prompt in point of time, then the oxygen inspired performs the kindest and most beautiful offices—the vivifying effects thereof being equally distributed through all parts of the dependent economy. Any deficiency of acid in the stomach and in the blood, or any excess of alkalinity and residuum in either, is sure to be succeeded by some local derangements in places the most impressible or predisposed. Palpitations of the heart are among the commonest symptoms of indigestion. In such case, we know that a rectification of the digestive functions would be *a cure*. Take a tea-spoonful of lemon juice just before breakfast and dinner—or, instead, eat a whole sour orange. Never eat between meals. Expand your breast, absorb the air, enlarge your spiritual heart, and all the rest shall be added—refreshment, strength, health, happiness.

Heart Disturbances and Nervousness.

Your heart-troubles are occasioned by magnetic exhaustion and dyspepsia. For long years you have swallowed enormous quantities of coffee every day. Your temperament is motive-mental; that is, you are overflowing with mental activity, and with nervousness contrary to all the laws of health and longevity. You are wearing out and wasting rapidly away. Now, put yourself under the discipline of the WILL. The effect should be motion only when motion is required by your occupation.

When you can, take Rest in Spirit—do not "fret," and "stew," and "worry" the life out of those about you. This will soon relieve the "heart disease." No patent medicine in America can furnish you with the elements of health.

Dyspeptic Pain about the Heart.

For breakfast eat plenty of brown or rye bread crumbled into a tumblerful of sour milk, in which you have first beaten up a fresh egg. Use very little sweet—no pies, no puddings, no drinking between meals. Twice a week, before dinner, dissolve fifteen grains of magnesia in a wine-glass of peppermint water, and drink it at once. Manipulate your waist, and knead the bowels while suffering from a dyspeptic pain in the heart. You may fortify yourself, if you will but obey the laws of life, against the fillibustering approaches of Gen. De Bility.

For Lump in the Throat.

Manipulate the throat from behind the ears forward, and thence down over the breast, just previous to retiring for the night. Once a week it will be well to anoint your whole body with good olive oil, mixed with a little essence of spearmint Restrain your desire for stimulations of either tea or coffee. May the angels sing their songs into the heart of earth's children. Be thou a justice-lover and a peace-maker in the land of thy fathers. The upper spheres will aid thee and thine

Canker on the Tonsils.

The simplest remedy for chronic affections of the throat, with canker-ulcers on the tonsils, and lassitude throughout the system, is this: Thoroughly rub the skin of the neck with cicuta ointment, such as is put up by druggists—*Unguenta Conii*. Then envelop the throat in a compress of lamb-skin or cat-fur. Remove it every morning, and bathe the parts in cold water, always drying by means of manipulations. Internally, the following is the best general remedy: Golden seal (*hydrasis Canadensis*,) and Pleurisy-root (*Asclepias tuberosa*,) of each four ounces. Make a quick fire, and boil these ingredients, in three pints of rain-water, down below one-half. Strain when cold; add one pound of brown sugar, and one tea-spoonful of red pepper. Dose.—Take a tea-spoonful in the mouth and swallow slowly, whenever the soreness is most severe, or when the body feels most like dropping with debility. Three or four doses per day may be taken with benefit.

Diptheria and Sore Throat.

Diptheria owes its origin to certain atmospheric influences, which are generated and widely diffused in some localities, while other and adjoining regions are wholly untainted by the poisonous vapor. Any one is liable to an attack of this subtile inflammation, and there is no condition, save that of sound health, which may be considered a harbinger of safety. Doctor Bruce says: "Upon a careful and somewhat extended investigation of the history of Diptheria, I find that it raged as an epidemic in Rome A. D. 380; Holland, in 1337; Spain, in 1600; Naples, in 1509; New York, in 1611 and 1771, when it was extremely fatal."

Symptoms.—The most prominent symptoms are that of weariness through the joints, and the sensation of a cold in the head, and throat, and lungs. Sometimes, however, the throat is not sensitive even when the diptheritic exudation has commenced. A fetid odor in the breath, and some slight redness and enlargement of one of the tonsils, are among the incipient symptoms.

Remedy.—Stop all food, even when your appetite is good, except gruels, porridges, and panadas. Drink not a spoonful of cold water. Bandage your entire throat, in early stages of the disease, with several folds of flannel. Keep this cravat on both day and night without changing. Be very quiet, and do not fear the progress of your disease. Gargle your mouth and throat every *half hour* with a strong gargle made of vinegar, honey, red pepper, and salt, mixed in a tumblerful of warm water. Do not go out of a warm room for several successive days. Breathe the vapor of hops occasionally; also sleep on a pillow filled with them. Take no physic or emetics. Keep the bowels open by warm water enemas. (This course, accompanied with some gentle magnetic passes to quiet the nervous excitement, will check almost every form of throat disease.)

Putrid and diptheritic inflammations of the throat, although resembling croup in many symptoms, should not be treated like the latter, but invariably as you would attempt to prevent an attack of *yellow fever*, viz.: By bathing the extremities in hot mustard water, rubbing them until the skin becomes very tender, and then enveloping them in many folds of flannel. The *American Medical Times* calls attention to the efficacy of creosote as a local application for diptheria. Ten drops of creosote to a gill of warm water is applied as a gargle; one or two applications effect a cure. Try it. Just balance the system in regard to temperature, give it plenty of rest for several successive days, and you will escape almost every form of putrid inflammations and eruptive fevers.

Malignant Sore Throat and Croup.

This semi-throat disease sometimes appears among adults, though children are most commonly subjects of it. It is a strange and painful malady, arising from atmospheric conditions, and should be promptly treated as an electrical affection. Its symptoms are somewhat like croup, combined with malignant scarlet fever, attended with occasional vomiting and purging, and concluding with the formation of a membrane or transparent film, which, covering the windpipe closely, brings on horrible sensations of choking and suffocation. Violent delirium is a very possible symptom in a certain advanced stage of the affliction.

REMEDY.—Treat the patient vigorously, as in a case of yellow fever. That is to say, bathe the arms and legs with warm water containing as much mustard or red pepper as the skin will bear without blistering. Or, which is a good substitute, bathe the arms and legs with camphorated alcohol and warm water; and keep the head very cool by constant application of cloths dipped in ice water. Then manipulate downward rapidly, until the surface is quite red and sensitive with the friction and irritation. *Stop all food*, and give only a tea-spoonful of ice-water at a time. Use frequent cold compresses upon the throat. Give from one to three warm water enemas each twenty-four hours. The wet-sheet pack is good when the surface is dry and hot. In extreme cases, where suffocation seems unavoidable, apply fresh beefsteak compresses to the throat. Rather than permit the disease to proceed, bathe the patient's extremities every half hour.

Sore Throat and Bitter Stomach.

For bronchial weakness, if long-continued, we prescribe the constant wearing of lamb-skin in front over the lungs; and at night, after severe symptoms, a bandage of the same about the throat. Keep the skin of your body clean, and get a fresh lamb-skin when necessary. Wash your whole person in soap and water once a week. For a sour and bitter stomach take a tea-spoonful of yeast three times a day; or, about twice a week, drink a bowl of weak tea made of equal quantities of wild cherry bark and spearmint, or take a three-grain pill after supper, made of equal proportions of extract of liverwort and dandelion. These medicines are prescribed to operate favorably in cases where the well-known common laws of Health are obeyed, and where stimulating beverages and physical excesses are avoided.

Bronchial Weakness, and Neuralgia.

As a general fact in this climate, a *Bronchial* weakness is a symptom of a low condition of the vital forces. The bodily powers drop below their true level, and then, in America, the symptom is either a neuralgic attack or a diseased condition of the throat. We allege the basic cause to be *hastiness, over-work, over-eating,* and *improper digestion,* and early neglect of physical functions, particularly the bowels.

REMEDY.—Give conscientious attention to the demands of the different functions; live regularly, systematically, in every external particular; and then, with great confidence, you may apply to your body, throat, back, and bowels, this ointment: Take tincture of arnica, olive-oil, oil of spearmint, and common turpentine, of each four ounces; pulverized camphor gum, a table-spoonful; prepared "goose grease," or refined lard, sufficient to reduce the whole, over a slow fire, to the consistency of molasses. This wash may be thinned by the introduction of alcohol. This *unguentum* will penetrate your venous system, and remotely affect the nervous forces, if you apply it, as directed, twice a week. Use it every night over your lungs and about your throat, when either are sore or debilitated.

Consumptive Irritation in the Throat.

A correspondent describes many cases in writing of his afflicted wife: "Last spring we were surprised to find that her lungs were very much diseased, according to the diagnosis of two physicians. She has taken no medicine, but has practiced your direction in regard to breathing, early rising (until recently,) and the Will-power. By the use of these natural means, I think the lungs are nearly or quite healed, strength very much improved, and general health vastly better. The disease, however, seems to be now in the bronchial tubes, producing cough, &c., and the above means do not seem to overcome it."

REMEDY.—Wear cotton stockings with fur insoles, and fur wristlets also, every night, or wrap the extremities in folds of flannel. Sleeping with these points *very warm* (magnetic,) will greatly aid Nature in the curative process. The cough will yield to something like the following : Pulverized pleurisy-root (*asclepias tuberosa,*) two ounces ; extract of dandelion, six ounces ; extract of hoarhound, the same quantity ; cayenne pepper, one tea-spoonful ; onion-juice, three table-spoonfuls ; brown sugar, one pound. Boil and mix, and convert this compound into *candy balls* about the size of white walnuts. DOSE.—Use this candy as freely as you can without nauseating the stomach —especially during parts of the day when the cough is most troublesome, or the throat the most sensitive. Do not abandon the pneumogastric efforts at Self-Healing.

Bronchitis, and Mucous in the Throat.

At night, and on rising in the morning, there seems to be an accumulation of mucous in the throat. After several attempts to clear away the obstruction, the patient finally *coughs* two or three times, which may bring it away in the shape of mucous and hard clots of yellow matter, sometimes streaked with blood. This may all occur in one who is temperate, eats no animal food, but leads a somewhat sedentary life.

REMEDY.—These symptoms are owing to a depression of the vital forces in the pneumogastrical department. No medicines can cure unless aided in their operation by regular exercises of the chest and throat, by means of breathing and reading, or vociferating with earnestness for a few minutes every day. We prescribe for incipient bronchitis: One drachm of nitric acid in two ounces of pure water. DOSE.—Three drops morning and night, on sugar. Occasionally take a vapor bath, and rub the whole body with equal parts of sweet oil, spearmint essence, and alcohol. Frictionize the skin with hair mittens.

Pulmonary Weakness and Irritation.

Many persons are threatened with a scrofulous irritation throughout the entire breathing organism. The membranes of the stomach are loaded with a gelatinous deposit from the vitiated condition of the lungs. And yet the lungs may not be diseased. A weakness is apt to spread over the nervous system, and react upon the spinal-centers, with some neuralgia.

REMEDY.—Bring the motor nerves under the control of Will. Exercise the stomach and bowels by exerting your Will upon the muscles that support and regulate their external interests. Your cure will begin with the just and appropriate performance of the digestive functions. Use the pneumogastric exertions immediately after getting in bed for the night. Breathe the fresh morning air freely, then swallow a few grains of red pepper mixed with Tolu syrup, or use it as a gargle. Place new wool, fur, or flannel, on any part of your body which is affected quickest by very cold weather, or by long rain storms. Obey the laws of life and health with respect to sleeping, eating, and drinking, otherwise you may very soon become incurably diseased.

Irritation in the Throat.

Many suffer from a continual irritation in the throat. We know that, with few exceptions, such persons are constant drinkers of coffee at the breakfast hour. And we must inform you that nothing can *cure* your throat while you continue the use of coffee as a beverage. Besides, no one can reasonably expect medicine to heal diseased membranes, unless the limbs are perfectly protected, and the extremities kept habitually dry, after duly bathing them. The climate of this continent will not justify any long-continued physical exposures—not even occasional dampness of the feet. No reader of this volume need plead *ignorance* of the laws of Health.

Naked Arms and Sore Throat.

The following wise words are taken from " Lewis's New Gymnastics." Let every mother read the truth, and then see whether the dresses of her little ones correspond: A distinguished physician, who died some years since in Paris, declared: " I believe that during the twenty-six years I have practiced my profession in this city, twenty thousand children have been carried to the cemeteries, a sacrifice to the absurd custom of exposing their arms naked."

I have often thought, if a mother were anxious to show the soft, white skin of her baby, and would cut out a round hole in the little thing's dress, just over the heart, and then carry it about for observation by the company, it would do very little harm. But to expose the baby's arms, members so far removed from the heart, and with such feeble circulation at best, is a most pernicious practice.

Put the bulb of a thermometer in a baby's mouth; the mercury rises to 99 degrees. Now carry the same bulb to its little hand; if the arms be bare, and the evening cool, the mercury will sink 40 degrees. Of course, all the blood which flows through these arms and hands must fall from 20 to 40 degrees below the temperature of the heart. Need I say that when these cold currents of blood flow back into the chest, the child's general vitality must be more or less compromised? And need I add that we ought not to be surprised at its frequently recurring affections of the lungs, throat, and stomach?

I have seen more than one child with habitual cough and hoarseness, or choking with mucous, entirely and permanently relieved by simply keeping its hands and arms warm. Every observing and progressive physician has daily opportunities to witness the same simple cure.

Swelling of the Throat.

Now and then a female will complain of a sense of weight, quite oppressive at times, on the throat and lungs. This symptom is caused by an internal swelling of the muscles and cords of the entire neck. The cause of such symptoms is inseparable from the reproductive organism.

REMEDY.—Use your Will-power on the intestinal regions. Lift the whole lower system upward with all your Will, particularly while walking, and breathe deeply two or three times per day. Get some friendly hands to squeeze and press your throat together, extending the same operation downward over the back and lungs. At night sprinkle a little gum myrrh and camphor between two layers of cotton, with which envelop the throat. Use cold water over your kidneys every morning.

Clergymen's Sore Throat.

A correspondent of the London *Times* writes: "Medical men recommend all public speakers who have a tendency to 'relaxed uvula,' 'clergymen's sore throat,' or 'aphonia clericorum,' to let the beard grow under the chin, and I cannot recall any one case of this complaint where this treatment was adopted, while all which I remember happened to clergymen either *beardless or shavers under the chin.* Again, when the exposure to wind and weather, to which the active clergyman submits in the discharge of his duties, day and night, is taken into consideration, I think no reasonable man or woman can refuse us the use of the protection which Providence has given us. The parson in the Yorkshire moors, the eastern county fens—loved home of piercing east winds—the wilds of Gloucestershire, or the chalky hills of the Southdowns, as he wends his way through lanes bounded by hedges no thicker than tooth brushes, or over country with no hedges at all, is victimizing himself, not to do his duty, but to the absurdities of fashion, and brings on himself all sorts of thoracic and pectoral woes every time he shaves. Possibly the Bishop of Rochester thinks a woolen comforter a sufficient substitute for Nature's own covering : but the fact is, that more colds are caught with a comforter than without. It chafes, and so produces an artificial heat, and often a great perspiration, and then a draught or chill comes, or the comforter gets loose, and a cold follows ; while, if the mouth is covered, the breath is driven inwards and damps the neckcloth, and so we get a sore throat. I heard from a laboring man, a week ago, a striking argument in favor of beards. He used to spit blood, and was in a bad state of pulmonary disease. Last spring he ceased to shave ; since then he has not only not spat blood, but also gained over thirty pounds of flesh in weight. How many consumptive clergymen might now be strong and useful, if they had had equal wisdom!"

Tuberculation and Adhesion of the Liver.

We have knowledge of a gentleman who has led a literary life for the past ten years. His reading has been considerable. Many hours of each day, during those years, have been consecrated to study and reflection. Physical inactivity, at times, was unavoidable. From various causes, a few small tubercles early formed on the right and upper edge of the left lobe of the liver. But his general health being good, and his temperament wiry and elastic, the parts partially healed over and ceased to give pain. The healing process, however, was not perfect. A somewhat raw surface was exposed, and the result is, *an adhesion* of a small portion of the liver to its investing membrane, a cellulo-vascular covering, which is intimately connected with the portal veins and hepatic arteries, with the ducts, and nerves, and absorbents, which are inseparable from the organ and the performance of its varied functions.

Symptoms.—A pain in the right side, extending from about the middle of the ribs down to the hip; sometimes the pain is only slight and confined to the lower part of the ribs, or rather to the region under the middle of the floating ribs; at other times it is quite bad, and extends up and down, and even to the left side. At all times, if he breathes freely or inspires deeply, he feels as if his side was kept back from a full expansion by some kind of compression.

Remedy.—Wear a bandage of thin india-rubber, six inches wide, about the waist and over the parts affected. It should be put on as tight as possible without painfully affecting the breathing, but only at night, being careful invariably to separate the rubber from the skin by the interposition of some light fabric, except, of course, the parts diseased, upon which the bandage should press close enough to exclude the air. After the treatment is persevered in for a number of nights, then practice deep inspirations every day, accompanied with muscular efforts to rub and thrash out the soreness. No person with tuberculated lungs or liver, either with or without adhesions, need fear any injury to arise from judicious *pounding and rubbing of the most painful parts*, aided by deep inhalations of pure air, in large quantities, through the nose.

Cure for Parasitical Croup.

In these days of diptheritic affections, by which many, both children and adults, are hastened prematurely into the transmundane sphere, the scientific treatment of croup may not be inappropriately considered. We know, by clairvoyant inspections, that the membrane of the throat will, when inflamed or thickened by cold, produce quite a crop of moss-like sores, fungi, which may be destroyed by the prompt administration of diluted nitric acid, and gargles of red pepper tea, sweetened with honey. "The Dublin Hospital *Gazette* states that Doctor Jodin, in a communication to the Academy of Sciences, on the nature of croup and on the treatment of the same, says that his researches have led him to the following conclusions: First. That croup and pseudo-membranous angina are merely parasitical diseases, due to the formation of fungi. Second. That the treatment of these affections requires neither general medication nor incendiary cauterizations, and that they may be cured by simple parasiticidal applications. After enumerating the various therapeutical means resorted to in this and analogous diseases, Mr. Jodin declares that he much prefers to those uncertain, alarming, or dangerous remedies, *the sesquichloride of iron*, which completely impregnates the fungus, exercises its action on the surface only, and may be absorbed without danger. This medicine destroys the parasitic growth, and also modifies favorably the hemorrhagic condition constantly observable in the affected parts and their neighborhood; it further induces expectoration, and thus promotes the rejection of the false membranes."

Remedy for Loss of Voice.

Aphonia may be caused from an inflammation of the parts around the larynx. But in most cases, the loss of voice is owing to a deficiency of tone, a weakness of the muscular fabric of the stomach and throat, a kind of *atony*, as the doctors express it, and the most distinct symptom is nervousness of the entire pneumogastric region.

REMEDY.—The general system must be strengthened. To do this it will be necessary to eat a moderate breakfast of brown bread, with, perhaps, one boiled egg; or two soft-boiled eggs, and nothing else. For drink, a small quantity of Brewer's ale, mixed with an egg, and sweetened to your taste. No salt meat for dinner. Fish will do, if not salt, and bread and butter. Nothing hearty for supper; no fruit, cakes, or sweetmeats. Bathe your entire chest in cold water every morning, followed with considerable friction, and bandage your throat with a cold compress every night. Always remove it in the morning, and shower the entire throat and shoulders with cold water. Chew occasionally a little sassafras bark.

Bronchocele, Goitre, or Big Neck.

The enlargement of the thyroid gland of the neck, which slowly assumes a firm fleshy appearance, extending toward the sides of the throat, and sometimes attaining to very large proportions, is caused by a variety of influences affecting the lymphatic glands and cellular tissues in early youth.

REMEDY.—This must consist of timely efforts, before the painless enlargement has assumed a fleshy hardness, otherwise the patient may not expect permanent relief. Let every person remember that the early "ounce of prevention" is a cure, when the "pound of cure" is worse than nothing. Our only general prescription is, go out every day to some running stream of fresh water, bare the neck, and bathe it long and thoroughly; then dry the skin by continuous manipulations, and cover the throat with a soft fur cravat.

Cure for Tuberculosis.

Taking as a basis the sum generally considered as that of the population of the globe, it is fair to estimate that from eighty to one hundred millions of its inhabitants succumb, by a premature death, to some form of this disease. It destroys nearly a sixth of the population of England. We prescribe *Breathing* as the remedy and preventive.

Objections to Deep Breathing.

A lady correspondent, who, for a time, adopted our *Deep-breathing Cure*, writes as follows : " Let me mention an objection which I find to the application of your prescriptions for *deep-breathing*. I have been experimenting in that way for a few weeks, and now have *not a garment that will meet around my waist*, owing to the expansion of my lungs ! Perhaps some of my Sisters may be deterred by the prospect of such inconvenience, from seeking health by this method.

"Faithfully yours, MATILDA."

ANSWER.—Many of Matilda's fashionable Sisters will undoubtedly be "deterred by the prospect of such inconvenience," and yet, somehow, we are credulous enough to believe that there are thousands of progress-loving mothers who will not only practice deep breathing, but will instruct their daughters to obey the laws of beauty and health in this respect, even if a "Reform dress" should become the eventual necessity. We shall see.

Enlargement of Glands in the Throat.

Your sore throat may be caused by a singular enlargement of the cartilage at the top of the swallow. It may be partly caused by nervous anxiety, or care and watchfulness, or by dreams. The true remedy, besides pure cold water, is magnetization by some congenial hand. The throat should be squeezed and softened by tender pressure, as you would mellow a peach. And, above all, you must soon forget that your throat is affected. If the glands are much swollen, use poultices of Life-everlasting every other night. Shower the neck every morning with plenty of cold water. This will stop the enlargement ; perhaps will cure it.

Remedy for Difficult Breathing.

There are several modes of treating such a negative condition of the lungs, but it is deemed wisdom, first, to reach and arouse your *energies* through the stomach and the skin. Perhaps the original cause of your present prostration was imperfect chylification. Your liver has been torpid, your kidneys much disabled, and the ultimate of all disturbances is fixed upon the lungs. It will be necessary for you to use weak chamomile-tea injections three or four times per week, for a month or two. The intestinal action must become natural and prompt. A tea-spoonful of pulverized prepared charcoal in a wine-glassful of lime-water, is important just before eating every other dinner. Arise! This life is yet best adapted to thy development. Use your Will-power! Swing your arms backward, upward, and walk with your mind in *your feet*. (This counsel is to every sick person.) After one week of this treatment, prepare a large Burgundy pitch-plaster, sprinkle the surface evenly with opium, and apply it all over the upper portion of your chest. When this wears off, you should either renew it or apply the treatment recommended for a "Negative Condition of the Lungs." Eat moderately at noon, with plenty of red pepper (*capsicum*,) on cooked or broiled meats.

Prevention of Cold-taking.

We would most gladly aid our Brothers and Sisters in the effort to keep an equilibrium of bodily temperature, which is the surest preventive of colds, catarrhs, bronchitis, &c., &c.; but the *Scientific American* has produced a paragraph so entirely to the point, that we cannot refrain from substituting it for what we were about to write on the subject: A "cold" is not necessarily the result of a high or low temperature. A person may go directly from a hot bath into a cold one, or into snow, even, and not take cold. On the contrary, he may take cold by pouring a couple of table-spoonfuls of water upon some part of his clothing, or by standing in a door, or before a stove, or sitting near a window or other opening, where one part of the body is colder than another. Let it be kept in mind that uniformity of temperature over the whole body is the first thing to be looked after. It is the unequal heat upon different parts of the body which produces colds, by disturbing the uniform circulation of the blood, which in turn induces congestion in some part. If you must keep a partially wet garment on, it would be as well, perhaps, to wet the whole of it uniformly. The feet are a great source of colds, on account of the variable temperature they are subjected to. Keep these always dry and warm, and avoid draughts of air, hot or cold, wet spots on the garments, and other direct causes of unequal temperature, and keep the system braced up by plenty of sleep, and the eschewing of debilitating food and drinks, and you will be proof against a cold and its results.

Coldness in the Back of the Neck.

A lady complains of a coldness on the back of her neck, which has troubled her for several years. In spite of many folds of flannel, there is always a cold *sensation*, except in the warmest weather.

CAUSE AND REMEDY.—The origin of a local chronic coldness is an inflammation of the membrane (*the periosteum*,) that covers the bone and over-taxes the fibers in the affected region. Liniments, and bandages, and plasters, are useless, except as palliatives and temporary protection. The parts are constantly exhausted of the vital magnetism. Electricity accumulates and escapes at the point of *coldness*, and yet all about that *negative* pole there is a circle of smothered inflammation. The remedy, therefore, consists in equalization of the vital currents. This course will reach all the membranes of the system, no matter how remote from the part affected. A sponge full of cold water squeezed upon the *cold spot* every morning, followed by vigorous rubbing of the adjacent parts, and pounding or chafing the stomach, waist, bowels, and awakening action generally in all parts of the legs and arms—this is the radical remedy. But it is also good, occasionally, to fix a small bag of table salt upon the coldest place. Equalize the general circulation, and your health will become sound as an angel's. This prescription will disappoint no patient.

Bleeding from the Nose.

The general cause is concealed in the nervous system. Debility of the nervo-power is the usual cause of all hemorrhages; whether from the lungs, anus, nose, stomach, or other parts of the body. And, therefore, the true remedy consists in whatever restores vigor to the nervous system. Witch-hazel bark, or borax, pulverized, applied within the nostril, will stop the bleeding very soon. Cold water should be poured on the wrists and nape of the neck, until the parts are very much reduced in temperature. Also, before the bleeding is likely to commence, press the large veins on either side of the throat, rubbing downward, thus arresting the extra determination of blood to the head. This should be remembered in sudden attacks of headache. Lift the nervous vigor by a gentle diet of fruits and grains. Animal food is unfavorable to a rapid up-building of the nerve-powers. In addition to the foregoing, we suggest, as conditions of cure, " temperance in all things," and plenty of " time."

The Theory of Hyper-Oxygenation.

A New York physician has started a new theory of tubercular consumption; the points of which are: "1st. That the morbid condition of the blood in this disease is one of Hyper-oxygenation, or an excessive amount of oxygen in proportion to its oxydizable material. 2d. That there is due to this oxygenation a certain pyæmic diathesis, or production of partly-formed pus in the blood. 3d. That tubercle is nothing more than this purulent formation excreted as effete or dead matter, and impacted in the air-cells of the lungs. The immediate cause of Tubercular Consumption being Hper-oxygenation of the blood, the indication of cure consists in the administration of an agent having a strong affinity for oxygen, thereby *acting as a preservative*, and preventing both the rapid oxydation of the tissues, which is *the cause of wasting*, and the formation of purulent matter in the blood, which is *the cause of tubercular deposit.*"

We think that a certain amount of truth lies at the foundation of the above theory, but we do not think that consumption, in the majority of cases, arises from a superabundance of oxygen, by which the tissues are consumed. Persons with weak digestion, who habitually violate the law of inspiration—that is, those who customarily eat too much and breathe too little—are the surest victims of consumption in this climate. More people die "for want of breath" than because of an excess of the oxygen in the air. The best remedy for consumption is a reasonable quantity of most agreeable food at each meal, a firm Will, systematic breathing before eating and after retiring for the night, and throwing the shoulders backward in order to expand the lungs, so that they may admit more life-giving oxygen with the magnetism of sunlight.

Persons who sit several hours per day, while at their labor, should inspire large quantities of air, thus filling every part of the lungs with the element of life. The shoulders should be thrown back, and the head kept erect, so that the smallest crevice of the pulmonary structure may be perfectly filled and expanded by the breath of heaven.

Negative Condition of the Lungs.

This condition is a cold state of the pulmonary membranes. The small vessels in the chest are swollen and incapable of discharging their duties. The symptoms are oppressive breathing, asthma, throbbing at the heart, and headache at night.

REMEDY.—In any case or climate the sweet-oil bath is first necessary—that is, anoint the back and body thoroughly with olive-oil and turpentine, warmed together and mixed by means of alcohol and a little of the oil of spearmint. Then apply a coating of raw wool, or cotton, over the entire back and lungs, including the throat, and cover it tightly with a suitable bandage. A temporary wool jacket might be constructed for this purpose; but it will be necessary to use fresh wool after wearing it a few days and nights. The effect of this bandage is legitimate and beneficial when the surface covered by it is heated almost to a sweating temperature. It should be removed every night, previous to going to bed, for the purpose of re-oiling the skin or cleansing off the perspiration. Then replace it as before, and sleep in it. Breathe! Breathe!

Negative Lungs and Cold Extremities.

A weak and cold condition of the lungs (by which the entire system is affected, giving cold extremities, and a frequent liability to taking "a cold,") may be more than half removed by covering the bosom and throat, for a few nights, with a light paste composed of dampened flour well sprinkled with pulverized camphor. Spread this camphorated paste on a thin cloth, and cover the outside with flannel. For sudden attacks of rheumatism, this remedy is very efficacious.

Cough and Incipient Consumption.

There is much bronchial weakness ; a loss of power in the upper part of the lungs ; some cough ; easy to take cold ; and premonitory signs of consumption.

REMEDY.—Practice breathing deeply every morning. Do not neglect to dress warmly. Walk from one to three miles every day. Get three quarts of good cider-brandy ; add one teaspoonful of powdered red pepper, three ounces extract of liquorice, four ounces of wild cherry bark. Let this tincture for three days. Take a wine-glass full early every morning. Stand erect! "*Blessed is the upright man.*" Straighten your spine. Expand your lungs. Throw back your shoulders. And breathe plenty of Heaven's pure ethereal flame. Come, reader! work with a good heart, support your family, think independently, let others enjoy the same luxury, and thus begin to unfold the Harmonial Man.

After Effects of Lung Fever.

Strengthen your lungs by Nature's only infallible medicine, namely, by a good digestion and by systematic discipline of the lungs. The after-effects of lung fever may be wholly removed by attention to food, dress, bathing, breathing, and exercise. Everything depends on your stomach. The best of food, with a feeble digestion, will ultimate in *depraved* blood ; and even so will the best digestion, supplied with unwholesome food, vitiate the crimson streams of life. "Blood food" is plausible enough in the light of chemistry; but, seen by the light of digestion and assimilation, it is supreme quackery. Every and any thing digestible by the stomach is "blood food." But it will not result in blood fit for circulation, unless the lungs furnish *pure air* in sufficient quantities. Give attention, therefore, to your stomach, and to the gradual expansion of your lungs ;

at least as much attention as you give to eating and drinking. Any straining of the pulmonary substance, by too deep inhalations, or by too rapid expirations of the air, will be considered unphilosophical and contrary to our counsel. But we advise systematic breathing, slowly in and more slowly out, for ten minutes three times per day. Avoid sudden changes of bodily temperature. Food and drinks should not be warmer than your blood. Take a vapor bath once or twice a week; always sponge off with cool water immediately after it. Be cheerful, faithful, beautiful, and progressive!

Treatment for Chronic Asthma.

Patients all speak of violent, crowing, suffocating cough, with constriction of the trachea; sensation as if there were too little air in the chest, with pain and pressure in the pit of the stomach, as if that region were too narrow; sleeplessness and debility; aching of all the limbs. Sometimes with flatulent colic and abdominal spasms; dry cough; contraction of the chest and larynx; have to sit up; relieved when coughing; suddenly wake in the night with dry cough. We think that, in true asthma, the immediate cause of the suffering is paralysis or falling of the floor of the lungs, viz.: the diaphragm.

REMEDY.—We deem this affection for the most part, like Fever and Ague, a periodical disease of the involuntary nervous system. The sympathetic ganglia, and the pneumogastric nerves, also, are many times principally involved. We insist on our Will Cure equally in every case of Ague and spasmodic Asthma. There are persons who can, if they will, stop the tremor of their system. The sure way, however, to defeat the object, is to believe *you can't do it*. But believe that *you* CAN; then we know you WILL. Several so-called remedies for Asthma are known to the medical fraternity. The fact is, there is no positive cure save in the pneumogastric nerves. Therein we find the *pharmacopœia* of all divine life, energy, wisdom, and health. Asthma is a *periodical* disorder of the pulmonary forces; sometimes it is an effect of other disturbances, and sometimes it exists and operates as a cause. But whether your asthma be an effect, or a cause, of other troubles, we urge you to treat it periodically—that is, do nothing and take nothing to avert the symptoms until a few hours or minutes previous to their reappearance. At such times put forth your strong Will; walk about *firmly, resolutely, determinedly;* and breathe slowly, but really, with *a steady might;* GET WELL.

Irritation of the Air Passages.

Irritation of the lungs, and much coughing, may be caused by *infusoria*, or little animals, which infest the air passages. They produce inflammatory conditions, plant the seeds of consumption, and sometimes ultimate in cutaneous diseases.

REMEDY.—Get some phosphorus, dissolve it in oil, stir it, and inhale the vapor, once a day. Also bathe the breast, throat, and spine, every night, with the following: Oils of amber and spearmint, of each half an ounce; sweet oil and laudanum, of each one ounce; alcohol, half pint; mix by shaking; apply plentifully with the hand, after having rubbed the skin with a dry towel.

Wasting of the Blood.

For wasting of the blood, which is an effect of long-continued disorders of the liver and stomach, accompanied by very great weakness and weariness, and some aching in the right side, with nausea and loss of appetite, we prescribe an ointment.

THE REMEDY.—Equal parts mutton-lard and gum of camphor—say eight ounces of each; first dissolve the lard over a hot fire; then stir in the gum as fast as it will melt and mix; and lastly, add a table-spoonful of red pepper, and thoroughly unite the mixture over the fire. Apply it cold, using your hands to all parts of the patient's throat, breast, back, sides, hips, legs, arms, feet, and hands, every morning, applying plenty of the ointment, and concluding by wrapping the feet and hands in folds of flannel. For food, use well-toasted rye bread with a few spoonfuls of chocolate. Whisky and linseed-oil, equal parts, a little at a time, will be useful. Three or four times a week swallow a tea-spoonful of powdered willow charcoal in a little gin, or sweet oil, and warm water. ☞ Use your hands magnetically while applying the ointment. The patient must Will to become whole.

The Duodenum and Sore Eyes.

Almost all cases of common sore eyes, with occasional inflammation, can be traced to a disordered state of the stomach and duodenum. Even if the food be properly digested, there is some derangement in the lower departments, where the bile joins the chyle, and where the ultimates of food are prepared, by the magnetic action of the mesenteric glands, for assimilation with the blood.

REMEDY.—About ninety minutes after eating dinner and supper, take a table-spoonful of the following tincture, as a promotive of secondary digestion: Mandrake, rhubarb, and red-pepper, of each (powdered,) one drachm; put them in one pint French brandy; let it stand a week; then add one pint of water, shake, and use as directed. If you travel, or cannot be at home at meal times, take a little bottle of this preparation with you. Wash your eyes in buttermilk whenever possible, or rub them with buckwheat flour. But without a radical change in your diet, the above medicines will avail you nothing. Do not eat meat at breakfast or supper; and not largely of any kind of animal food at any middle meal. No pastries of any description at any time; no fruit or vegetables for supper. "Under all cir cumstances keep an even mind." We will give every tobacco chewer a great task to perform for his own good—*Abandon the use of Tobacco in every form, and pledge yourself, in heaven's name, never to use it again!*

Prevention of Scarlet Fever.

If you suspect that your child, or any member of your family, is about to have scarlet fever, the best preventive is: Give three drops of belladonna, in a wine-glass of water, three times a day. Burn coffee in the room twice per day, and treat magnetically according to directions already given. Do not let the patient go out of the room until the inflammation of the tonsils and the fever symptoms subside, and then only when the earth is dry and the sun warm. Gargle with salt and vinegar in warm water, and sleep with the head more elevated than usual. If the bowels are costive, and the system is very feeble, give a tea-spoonful of castor-oil in as much good brandy, with a little water. Wash the body with some cooling fluid, or anoint with sweet oil, and manipulate.

Dry and Humid Asthma.

The only sure course to adopt in the treatment of this troublesome disease, is to apply your remedy just before a recurrence of the constrictions and paroxysm. The pneumogastric nerve is implicated; in fact, the disarrangement of this nerve is, in nine-tenths of cases, the cause of the disease. The *dry* asthma is quite unlike the *humid* affection; and the paroxysmal asthma is different from either; but the fundamental methods of the periodical treatment should be uniform—namely: When the symptoms are slight, give the whole surface a thorough oiling (1 tea-spoonful of Cajaput oil in 1 pint of Olive do., mixed with sufficient alcohol to unite the oils,) and immediately practice respiration, full and slow, expanding the bronchial tubes by throwing the head and shoulders backward. Use your Will. Wear a jacket next to your body made of oiled silk. Keep the skin clean as possible. When the breathing is most difficult, inhale a little, now and then, of the vapor of saltpeter.

The Wheezing, or Hay Asthma.

Pathological authorities will, no doubt, hesitate to believe that this species of asthma can be reached through the spine and kidneys. Let us try. Infuse two ounces of black henbane (*hyos cyamus niger,*) in one quart of brandy. Whenever the attack threatens to be severe, apply an onion poultice, saturated with this tincture, to the back, opposite the stomach. We think a few applications of this remedy, in connection with the breathing effort, will accomplish a cure.

Hemorrhage of the Lungs.

Adopt our principles of Self-Healing, and begin the work of Personal Reform from this hour. Get the resinous exudation of a pine tree, one drachm; the same quantity of cayenne pepper; make into pills of two grains each. DOSE.—One pill at ten o'clock each day, or immediately after considerable bodily exercise, and one the last thing at night. Take, occasionally, a tea-spoonful of olive oil in a little cider-brandy. Eat nothing sweet, drink nothing over-heating, but "be thou whole."

Treatment for Catarrh and Cough.

Many persons are afflicted with catarrhal symptoms; offensive breath at times; take cold easy; expectorate hard and dry mucous; and have a hacking cough.

Abolish the use of all sweet food and drinks; may eat plentifully of eggs, if not hard-boiled or fried in swine's grease;

no meat of any kind oftener than twice a week. The only two medicines we can discern are, first, the homeopathic preparation of *Phosphorus*, three drops in a wine-glass of water, whenever the symptoms are severe or troublesome, two hours between doses; and second, a lemon, slightly sweetened, as medicine every evening, between supper and bedtime. As soon as the symptoms subside, use the lemon-juice now and then, but only during the early part of the day. The bowels and digestive system generally must be kept healthy and *prompt*, by means of kneading and vigorous manipulations. No cathartics of any kind should be administered. These directions will cover a multitude of catarrhal cases, unless the patient is suffering from several diseases combined.

Soreness in the Lungs.

With a just and systematic treatment of the lungs, in the inception of their disturbance, all future trouble may be avoided. Watch the period of greatest uneasiness. When are you most conscious of having lungs? At what hour is the disturbance most distinctly realized? Apply your treatment accordingly. In going from the house into the open air, or the reverse, it is wisdom to keep the mouth firmly closed, using the nose for heating the atmosphere before it reaches the lungs. All persons, whether sick or well, should adopt this rule.

Sulphuric Ether for Deafness.

The new cure for deafness, which has created so much excitement among the physicians of France, has long been well known in this country. It consists in dropping sulphuric ether into the aural conduit. Several physicians in this city and elsewhere have, for years, employed it either alone or in combination with glycerine. By the latest accounts from Paris, we see that the opinion was beginning to gain ground, that the virtues of this remedy, though great, had been overrated.

Origin and Use of the Probang.

The probang was invented by a surgeon, to push down into the stomach substances which lodge in the throat or œsophagus. It is, properly speaking, a surgical instrument. Of late, it has been used to touch ulcers in the trachea, a dangerous method, which should not be adopted. A flexible piece of whalebone, with an oval piece of ivory, or a piece of sponge, fixed to the end, is called a probang.

Remedy for St. Vitus' Dance.

The best and simplest remedy for St. Vitus' Dance, in the young, is cold water and human magnetism. The water should be applied to the whole body every day, by means of a wet sheet and a plunge-bath afterward, and the magnetism by means of another's hands rubbing the body entirely and rapidly dry. Passes should be made from the back and sides of the neck to the ends of fingers and toes. Take mandrake-root, two ounces; lobelia leaves, half an ounce; boil in half gallon of water to one pint. When cold, add one-half pint of brandy and half pound of sugar. Dose.—One tea-spoonful before breakfast and dinner. Anoint the body with sweet oil.

Pain in the Joints.

In cases where the joints are painful and stiff, and particularly where the bones of the neck and back head are sore and rheumatically affected, we prescribe the following palliative, which may be applied with the hand: Common brandy, two gills; laudanum, one ounce; oil spearmint, one drachm; tincture arnica, two ounces. Mix, and use whenever the pain and rheumatic achings are troublesome.

Cure for Ill Temper in Children.

A sensible woman, the mother of a young family, taught her children, from the earliest childhood, to consider ill-humor as a disorder, which was to be cured by physic. Accordingly, she had always small doses ready, and the little patients, whenever it was thought needful, took rhubarb for the crossness. No punishment was required. Peevishness or ill temper and rhubarb were associated in their minds always as cause and effect.

Effect of Disease on Life.

By calculation, it is shown that, of 1,000 individuals, 23 die in their birth; 277 from teething, convulsions, and worms; 7 in measles; 2 women in childbirth; 195 of consumption, asthma, and other chronic complaints; 250 of fever; 12 of apoplexy; and 41 of dropsy. Or, in another point of view, of 1,000 persons, 200 die within the first year, 80 in the second, 40 in the third, and 24 in the fourth; and within the first eight years of life 445, or almost one-half of the number, are cut off by premature death.

Cure for Melancholy and Meanness.

A correspondent writes as follows: "I should be much obliged for a remedy for the following complaint: Dullness of mind; incapacity to study much without becoming sleepy; confused thoughts; some melancholy; and, perhaps, some meanness."

REMEDY.—It will be necessary, first of all, to wash your body every morning in cold water. Next, eat nothing animal for your breakfast, not even a grain of butter, nor drink a spoonful of milk. Next, eat a light dinner on the day when you wish to read and meditate. Never attempt to read or think soon after a hearty meal. Next, take up some one subject and concentrate your Will upon it, and think steadily while your Will is positive. For "meanness," take plenty of exercise, and think of the Summer Land. Let your daily life be regulated by fraternal love, Justice to thy neighbor, and the principles of untrammeled Freedom.

Headache or Cephalic Pills.

Cephalic Pills are good for periodical and nervous headache, chronic lameness, indigestion, nervous debility, &c., &c. Red pepper, whitewood bark, mandrake, life root (*Senecio aureus*,) of each, pulverized, half a drachm. Divide into pills of three grains. DOSE.—From one to four of these pills at night, or morning, when symptoms are severe. But there is *one secret* worth remembering, namely: That absolute rest to the stomach is one of the simplest means of cure, in both acute and dyspeptic diseases. Not one atom of food of any kind should be taken in any case of acute disease. Starve for twenty-four hours, and drink a little water; nothing further is necessary.

Cure for Cold in the Head.

Some individuals are constantly taking cold in the head, and having an ear-ache, almost every winter night, in spite of every precaution. The explanation of the ear-ache and head-cold, is: You sleep on a soft pillow, which allows your head to sink down into a hollow of feathers, by which considerable perspiration is promoted about your neck and ears; then, as soon as you change your position in bed, the cold air strikes the sweating surfaces, which very rapidly become negative, and hence the cold in the head and the distressing neuralgia. Of course you will hereafter use a pillow so hard that your head cannot sink into a heated valley.

Odor from the Nostrils.

The accompanying symptoms are: A sense of fullness in the head; of weight between the eyes; frequent sneezing; some hoarseness of the voice; occasional expectoration; and a disgorgement of fluid and *fetid* mucous from the throat. These symptoms are developed as the accompaniment of indigestion, fever, worms, measles, rheumatism, &c.

REMEDY.—The diet is the essential. Use no salt butter; no meat or gravies; nothing sweet after dinner. Milk is particularly unfavorable unless it is sour; but wheaten grits, rice, hominy, and Indian cakes and puddings, with stewed fruit, are useful and advisable. As a special remedy we recommend a cold water compress about the neck at night, and showering the head and neck behind the ears with cold water every morning. This should be done before washing the face. For the nose: Get of camphor twenty grains; blood-root half drachm; both must be perfectly pulverized, and mixed equally together. Use a very little of this powder, as a snuff, twice or thrice per day.

Cultivation of the Sense of Smell.

There is hardly any limit to the teachable capacities of the sympathetic nerves. The five senses—touch, taste, smell, hearing, and vision—are but different organizations or depots of the sympathetic ganglia, which are inseparable from the cerebral functions. Brain and nerves must be in a healthy condition, otherwise it will be impossible to greatly enhance the delights of this sense. The nerves of smell are spread upon the membranes that line the air passages. Consequently the *tasting* of an odor is as natural as the *feeling* of a sound. It is well known that deaf persons hear and enjoy the music of an orchestra. The intuition of cultivated smell is infallible. The North American Indians, before their overthrow and banishment by our ancestors, could distinguish different tribes, or the different members of the same nation, by the odoriferous exhalations of their bodies. Dogs, foxes, horses, and many other animals, are endowed with remarkably acute powers of smelling. Snufftakers are deprived of the delicate pleasures of this sense. Diseased and morbid nerves are keenly sensible only to odors the most repulsive. We know of mediums who first communicated with their familiar spirits by means of smelling sweet and aromatic odors diffused through the room. Several orthodox clergymen are said to have smelt—"brimstone!" We are strongly inclined to admit their testimony.

Cure for Common Deafness.

An aged man must not expect to keep his senses in full force. Many, however, suffer the misfortune of deafness to creep upon them from inattention to the ears during a cold or after a fever. In all cases where the wax is dry, the tympanum sound, but hearing imperfect, we have prescribed, with curative effect, the following : Nitric acid, one drachm ; water, one ounce ; phosphorus, one grain ; mix, and filter through a fine cloth, into a table-spoonful of burnt olive oil. Keep it tightly corked. Dose.—One drop in each ear (or in the one affected,) every morning.

If deafness arises from a recent cold, or exposure and fatigue, the patient can get relief by smoking dried hops, and then forcing the vapor from his mouth into the throat-passages leading to the ear-chambers.

Again, if your deafness is not a symptom or effect of some derangement of the system, but is traceable to a dryness of the natural moisture within the eustachian tube, then there is no simple remedy better than a draught of roasted onion in each ear. Or, soft clay made into convenient shape and size, then dampened with saliva, and put in the orifice of each ear, confining them with cotton and a bandage. All such applications are best made at night.

Ulceration of the Ear.

A remedy for chronic ulceration of the ear may be found in the common " Witch-hazel " (*Hamamelis Virginica*,) which is one of the most healthful productions of earth. Make a very strong decoction of all the fine parts of the bark scraped from the body of the tree. When done, you should not have more than a wine-glass full; add one drachm of borax, a table-spoonful of cream, and it will be ready for use in twenty-four hours. Saturate cotton with it, and stop the ears every night. ☞ In the morning, bathe the neck, behind the ears, and wash the orifices quite clean, before wetting the face and forehead.

Decaying Gums and Loose Teeth.

One preparation is applicable to the restoration of emaciated gums, namely: White oak bark, pulverized, one ounce; one ounce of camphor gum; quarter ounce of lobelia leaves, put together and tinctured one week in old cider-brandy. Clean the teeth and gums with pure soap, using a soft brush, and then saturate the gums with the clear tincture. This process should be adopted immediately after eating each meal. Breathe through your nose as much as possible, because the cold air, passing over the teeth and gums, tends strongly to their decomposition.

Enlargement of the Thyroid Gland.

An indolent enlargement of the *thyroid gland* is termed "Bronchocele." The incipient stages of this painless tumor are quite controllable by the persevering use of the electro-magnetic battery in connection with the hand. When it has existed for some time, and the soft elasticity of the gland has disappeared in a firmer enlargement, which extends toward the sides of the neck and involves the lymphatic glands, then the magnetic current is of little benefit. Iodine was long administered internally for this disease, with next to no benefit, but certainly leaving traces of its inadaptedness. In advanced stages of this swelling, we would recommend the frequent application of the magnetizer's hand. Let no patient expect an absorption of the gathered material after the hardening process is complete, unless by a most constant use of human magnetism, in connection with an occasional bathing of the parts with cajaput oil mixed with chloroform and bay rum.

Treatment for Old Sores.

The best general remedy within our knowledge for a disease in the feet and ankles, with sores that open and close for months, is a wash made of pokeberry leaves and roots, or of crab-apple tree bark, or of black Indian hemp. Make a strong tea of either, and use it to wash and cleanse the parts, with or without soap, as you feel inclined. It will greatly expedite the cure to expose the parts to the breathings of a horse, dog, or cow. If possible, let a dog sleep near, and exhale his breath over, the sorest surfaces.

Fever Sores and Tumors.

Syringe the sore every morning with the following liquid: White oak bark (inside,) four ounces, boiled in a quart of water to a pint. When cold, bottle it, and add one ounce of borax, in small pieces. Use a little for each dressing of the sore.

Bathe a growing tumor with the fresh juice of burdock. Pound and press the leaves until you obtain sufficient fluid to thoroughly saturate the diseased part. After this application, use the magnetism of your own hand; always rubbing from the top downward. Bathe the irritated surfaces thoroughly with pure soap and flax-seed tea, forming a suds, then take fresh slaked lime and dust the diseased parts at least twice a day.

Medicine for Scrofula and Erysipelas.

Let no patient expect to be healed by the performance of miracles. "Vicarious atonements" are dangerous in theory, and impossible in practice. You must individually *deserve* the possession of every luxury. If, therefore, you would be healed of eruptive diseases, itch, salt rheum, &c., let your table-habits be righteous and reasonable. Cancerous and scrofulous constitutions cannot be too careful about the solids and fluids which they eat and drink. In addition to directions which we have been impressed to give in a former volume, we will suggest the following beverage for scrofulous individuals: Mountain dittany, yellow dock, elecampane, and comfrey roots, of each eight ounces; white-pine, wild-cherry, and butternut barks, of each five ounces; guaiac chips, blue flag, and licorice-root, of each two ounces; break these ingredients together, put them into an iron vessel, and tincture them one week in three quarts of brandy, when strain off the brandy; then add three gallons of water, and steep the mixture over a slow fire for one whole day; add the same quantity of water, and boil the whole rapidly down to nearly one gallon; then strain it immediately. After this liquid is perfectly cold, add half an ounce of the muriated tincture of iron, return the brandy, and one quart more of good French brandy, and bottle it very tight.

This preparation may be regarded as wine. It may be diluted with water, and sweetened with sugar, as the patient's taste and strength demand. Whenever thirst is experienced, and while eating such articles of food as are prescribed by

30

experience, this pleasant and highly potent beverage may be used as wine. It may be much diluted—even a table-spoonful of this wine to a gill or two gills of water—and it will assist the diseased structures to harmony.

This *beverage* should be made and taken for at least one year, with occasional intermissions. It is designed for those various cutaneous conditions, already named, which are the incipient manifestations of the cancerous or scrofulous tendencies of the system; but it is particularly good, as a remedial drink, for such constitutions as are already suffering with the cancerous or scrofulous formations.

Treatment for Chronic Erysipelas.

This disease is characterized by a redness and inflammation of the skin at the place attacked, accompanied in a few hours by considerable swelling and oppressive fullness.

REMEDY.—The patient should have a vapor-bath, producing a thorough sweat, every day—perhaps twice, if the symptoms are violent—until the inflammation is all drawn out of the head, and a balance of the system is established. If the limbs are affected with the erysipelas fire, we still urge a complete sweating of the body every day, with proper attention to diet. The patient should go from the bath to the bed, and remain well covered for half an hour. Magnetic passes should then be made from the head downwards, over the hands and feet. Drink a tumbler of buttermilk every day. Diet should not be drawn from either the fruit or animal kingdoms. The various grains, made into puddings, and bread, and porridges, are best. The disease is not dangerous unless it is rapidly developed in the face and head—in which case, no remedy is comparable to human magnetism.

Medicine for a Canker Humor.

Thousands suffer from a depravation of the fluids which are propelled through the lymphatic system of glands and vessels. Nearly all cankerous and scrofulous forms of disease originate, first, in faulty digestion, and, second, in the vitiation of the functions of the lymphatic system. The remedy, in all cases, must be directed primarily to the nutritive functions, and indirectly to the lymphatic vessels throughout the body. The most convenient preparation within our knowledge is this: Scullcap (*scutelaria lateriflora*,) four ounces; spikenard (*aralia racemosa*,) six ounces; gentian (*triosteum perfoliatum*,) two ounces; cloves, pulverized, one table-spoonful. Put these articles in a jug containing one quart of water and one pound of sugar; cork tight, and let it stand near a fire for three or four days; then add one quart of best brandy, and you have a medicine which will penetrate and vivify the entire lymphatic system. DOSE.— Begin by taking less than a tea-spoonful about two hours after both dinner and tea. Increase a little every day until you can bear a table-spoonful; then add some more brandy and water to the medicine, and white sugar, as in the first instance. Eat and drink whatever suits you best. Arise! be cheerful!

An Intermittent Eruption.

The disease is entirely one of the under skin, appearing and disappearing on different parts of the body, and it is, therefore, absurd to drink blood-purifying syrups as a remedy. The best treatment for a simple incorrigible eruption, is a wash composed of kino, borax, and cream. Of borax take two ounces; two drachms of kino; cover them with one pint of rain-water, for three days; then add half pint fresh cream, and the mixture is ready for application. First give yourself a thorough scratchin or chafing, so that the surfaces are as much inflamed as they ever appear after rubbing them; then apply the mixture thoroughly. The smarting, burning, and inflammatory effects that succeed, will be of short duration, while a healing process will be forthwith inaugurated. Perhaps it will be necessary to repeat this treatment every night for a week or two, but the remedy is quite certain to remove the cause just beneath the skin. This treatment, remember, is adapted to simple eruptions, with inflammatory aspect.

Cause and Cure of Ringworm.

The technical term for this affliction is *Nerpes circinatus*, but physicians are not agreed as to the cause of this troublesome vesicular eruption. We have carefully observed that the circular patches float upon a transparent fluid; which, indeed, is *the cause* of the irritation and exfoliation—a saline acid, distilled from a diseased condition of the mucous membrane of the digestive organs.

REMEDY.—The pathological base of the ringworm is an erysipelous condition of the fluids and blood. Salt, therefore, must be avoided. Reject all food in which the saline property is prominent. Follow this plan throughout your earthly life; else you will never be free from the evil seeds of erysipelas. This rigid course will strike at the bottom of the ringworm diathesis. Bathe the eruption with half ounce of borax, dissolved in two gills of water, and one do. of cream. Add to this wash about fifty drops of sulphuric acid. Use it twice or thrice per diem.

Use of Sulphur for the Itch.

This affection of the skin has been displaced by the morbid conditions of erysipelas. It is said that "since the improvement in microscopes, the Itch is found to be a living creature burrowing in the flesh. This was discovered in 1812, by M Gales, apothecary to the hospital of St. Louis, in Paris. Experiments were immediately commenced to ascertain the best remedies to destroy so formidable an enemy. It was found that the Itch will live in clear water four hours ; in sea water it died in three hours ; in Goulard's solution it survived but one hour ; in castor, olive, almond and sweet oil, it died in two hours ; in croton oil in four hours ; in lime-water three-quarters of an hour ; in vinegar twenty minutes ; in alcohol twenty minutes ; in spirits of turpentine nine minutes ; in hydridate of potassium six minutes ; in solution of arsenic four minutes ; in sulphuric acid three minutes ; in creosote it died instantly ; placed on powdered sulphur it was found dead next day. From all these experiments, it was found that sulphur was the surest and safest remedy." Such is understood to be the history of the discovery that "Brimstone" (of the orthodox stamp,) is useful for the above disgusting eruption. Our impression is distinct that the "living creature" is an effect—not a producing cause—of the cutaneous affection.

Juvenile Excitability and Eruption.

The natural cure for youthful nervous excitability consists in a speedy development of the Nutritive and Muscular temperaments. To begin this work, it will be necessary to keep from the young stomach all exciting drinks and heating diets. No animal matter, no butter after breakfast, no milk after dinner ; but plenty of rice, barley, wheat (coarse,) corn, and sometimes rye. Simple puddings or cakes of these grains, with fruit and berries, in their season, constitute a fine diet for the development of young bodies. Much of the nervous activity will depart with the incoming of womanhood.

For the eruption we prescribe a strong wash of burdock (*arctium lappa*) roots, leaves, and seeds, to be used freely on the irritated surfaces once or twice per week. Many kinds of prickly eruption can be overcome by a plentiful use of burdock tea externally. When other remedies fail, try this in good faith.

Cure for Prurigo Pruritis.

In papulous eruptions, characterized at first by soft and smooth elevations of the cuticle, the surface-skin becomes diseased. Hence, when not cured in youth, or successfully treated when it makes its appearance, the disease slowly establishes itself in chronic form. It is now exceedingly difficult of management. The most convenient remedy is: *Equal parts spirits turpentine, and oil of sassafras, and sweet oil, amalgamated by alcohol.* Rub this thoroughly into the skin wherever the symptoms are exceedingly troublesome. If the tingling and stinging is too intense, add a coating of light cream. Dieting will make no difference. Avoid the extravagant use of salt.

A Nervous Rash.

The condition of many is described by a lady patient thus: " I have for about eight years been afflicted with a rash, appearing with great violence, whenever I am warm or excited—causing my head to throb violently, and my face to swell. The spells last usually about ten minutes, and then leave me shivering."

REMEDY.—Get Valerian root, half an ounce; extract of dandelion, one ounce; cayenne pepper, one drachm; cinnamon bark, one tea-spoonful—pound and mix these together; let them tincture in half pint of pure alcohol for ten days, then add one pint of pure water. DOSE.—Thirty drops in a wine-glass of water immediately after an attack, or when the "shivering" sensation pervades the nervous system. Eat no salt food. Drink little of any fluid, except when thirsty.

Small White Blisters on Children.

Children afflicted with any form of erysipelas, scratches, white blisters, sores, or scrofulous swellings, should never sleep with each other, never with a healthy child, and certainly never with adult persons. The magnetism of a spaniel, or half-cur dog, is healthful for children so diseased; and it would be beneficial to give a scrofulous child a dog (not a poodle or King Charley,) for a sleeping companion. A fresh cat's skin (*minus* the cat,) bandaged over the white swelling, or a piece of fresh beef applied every night to a scrofulous tumor or abscess, will effectually dissipate poisonous vapors, and partially restore the conditions of health. Diet and hand magnetism are the next best remedies.

A Virus Developed by a Scratch.

By particular inspection, it became evident, in the case of a young woman, that, in consequence of an unfortunate brier-stab in the superior nerve of her foot, a scrofulous and neuralgic *virus* had been diffused (rather aroused,) throughout her system. Sometimes it happens that a pin-scratch, or a needle-wound, at first inappreciable, will give rise to years of discord and suffering. You know how small a variation from the tone of the master-key will *untune* all the cords of the finest and most costly instrument of music—or, how slight an injury done the works of the best-made chronometer—even a grain of sand lodged in the wheels of your watch—will throw the movements out of harmony with the noiseless march of Time. O man! O woman! bow down and worship the Divine Wisdom—the great, beautiful, gracefully operative Principles of the Universe —from whose loving bosom proceeded the "fearful and wonderful" dynamics of individual existence.

The best remedial course to over-reach and remove the

nerve-virus will be : A thorough steam bath once or twice a week, about middle of the day. Food, for that day, gruel or porridge. Occasionally bathe the locality of the wound, and all the implicated parts of the limb, with a weak solution of ammonia, and a little chloroform. Use no salted nourishment, but take regular out-door exercise.

Cure for Itch and Tetter.

We have received upwards of one hundred applications from persons suffering with various eruptive diseases—Itch, tetter, rash, barbers' itch, salt rheum, &c. REMEDY IN GENERAL.—Eat nothing in the shape of swine's flesh ; abandon all salt foods and black pepper ; soups are good, but gravies injurious ; eat no butter after breakfast ; drink plentifully of buttermilk ; and cautiously bathe the affected parts with the following: Olive oil, half pint ; oil of spearmint, two ounces ; crude oil of turpentine, one tea-spoonful ; hartshorn, ten grains ; stir these into the well-amalgamated whites of six eggs. Always shake before applying it to the irritated skin.

Pityriasis, or Red and Rough Eruption.

The skin is at first only red and rough, but soon becomes branlike, or meally and scurfy. After the scales repeatedly form and separate, the exfoliations become rapid and troublesome. REMEDY.—Take equal parts of Witch-hazel (*Hamamelis Virginica,*) and butternut bark (*Juglans Cineria*) ; boil them with fresh lard down to the consistency of an ointment. Use it once a day. Eat plenty of tomatoes ; no animal food ; and abstain from salt.

Cause and Cure of Freckles.

Scientific investigators have given the term *ephelis* as a name for freckles, and very properly, we think, because it is clear to a demonstration, that (as the term implies,) the direct rays of the sun produce freckles, large brown patches, and other dusky blemishes, which appear on the face of persons of certain temperaments, who are not in perfect health. In a peculiar manner the skin is *oxydized* in spots, and these spots are freckles, or *ephelis*, as the ancients judiciously named them. The *remedy* will be found in phosphorus and acetic acid. Twelve grains of phosphorus in one ounce of the acid. First dissolve the phosphorus in as little olive oil as is necessary, then add the acetic acid. Keep this preparation tightly corked. First give the skin a slight coating of milk, then apply the preparation. Great care is needed, in making the application, not to get any in the mouth or eyes. Immediate exposure to the sun would prove unfavorable. One application a day is sufficient.

Worms Under the Skin.

Flavius, a student, writes: "For a year I have felt a prickling, burning, very disagreeable, and often painful sensation, a little above the outer extremity of the right eyebrow, thence upwards to the hair. Continued and difficult study makes it worse; sometimes, however, it becomes worse without any apparent cause, often resulting in a breaking out at the upper part of the affected region."

CAUSE.—A diseased action or obstruction, whereby are generated the depraved elements of the *scabies vernicularis*, but the healthiness of the system, and the activity of the nerves in that region, will prevent any formation except something in the shape of a sore, if allowed to proceed. REMEDY.—Bind on fine salt moistened with spirits of turpentine. Use the electromagnetic battery a few times.

Chapped Face, Hands, and Lips.

Druggists will charge from twenty-five to fifty cents for a "camphorated" preparation, which the reader may provide for a few pennies and a little industry. In cold weather, and particularly in windy seasons, when the delicate skin of the lips of unhealthy persons crack open and bleed, and the hands of outdoor laborers become rough and sore, it is always wise to have a remedy within the reach of all.

REMEDY.—Take equal quantities in weight (much or little as you please,) of fresh mutton-tallow and camphor-gum. Put them in a tin vessel, over a moderate fire, and let them melt and blend into one compound. Stirring the mixture will more perfectly interblend the two substances. This preparation you may term "camphor ice," or "camphor cake," or "camphor ball," or "armandine," or any other name more elaborate or chemically mysterious. Keep it in a convenient jar, and use it as often and as plentifully as your " chapped " surfaces may require.

Scar on the Face.

We do not perceive any cure for the " scar " on your face. Better be thankful that it is not a disease, and in your joy forget the blemish through all this life. The next time you are born, the form and features of a beautiful face will be revealed, free from all earthly accident, spot, or blemish. Some excellent souls thus wait the inheritance of their fortunes. But *real* beauty, after all, comes from pure spiritual affections and harmonious states of mind. With these elements behind the physical, within the bosom and within the brain, the external, however deformed, will take on the light and graces of beauty. Without such interior elements, the most attractive form and face slowly waste and crumble into hideous deformities.

Voluntary defacements should be prohibited. The dangers of tattooing, so much practiced among seamen, have been pointed out in a recent report of the Inspector General of Health of the French marine. The loss of an arm, and even death itself, are shown to have resulted from operations of this nature, while minor accidents from this cause are very numerous. Authority calls on sailors, inviting them to abstain from this dangerous practice.

Cancer cured by Nitro-muriatic Acid.

A jeweler, who had a bad cancerous pimple on his cheek, having occasion to dissolve some gold in nitro-muriatic acid, rubbed it several times, unconsciously, with his impregnated fingers, and was surprised to find it speedily change its appearance, and shortly disappear. M. Recamier, suspecting the cause, made several uniformly successful experiments of the same mixture, and thus by accident discovered a new caustic for cancerous affections. The proportions are one ounce of the acid to six grains of chloruret of gold.

Confessions of a Cancer Doctor.

Most of the vegetable plasters for cancer owe their activity to mineral substances. One consists of arsenic and sulphur, with powdered crow-foot leaves mixed with white of eggs; another of arsenic and extract of conium; another of sulphate of zinc and extract of blood-root; another of chloride of zinc and blood-root; another of potash alone, boiled down to the consistence of a plaster; another is made by evaporating the spirituous infusion of bitter-sweet, stramonium, conium, belladonna, yellow-dock, and poke, adding fresh butter to make an ointment; the poke and dock-root are also used separately. A celebrated cancer-powder is composed of arsenic, charcoal, and cinnabar. The acetate of copper mixed with vegetable extracts is also used. The chloride of zinc, however, is now the remedy most relied upon, and generally it is mixed with the extract of bloodroot. Many obstinate tumors and ulcers have been cured by the above remedies: all of which are called cancers, but few of them are really such. The last-named is mainly used to kill tumors, and make them fall out of their places, which has been practiced for a hundred years.

Nature and Cure of Scarlet Fever.

Several mothers have written for remedies to cure their little ones of scarlet fever. This disease, sometimes termed "Scarlatina," from the Latin *scarlatto* (meaning a deep red,) is a kind of spring contagion among children. Measles, mumps, chicken-pox, and the like, so common to childhood and youth, may all be prevented. They are caused mainly by the accumulated matters and vapors in the blood—a result of stimulating diet during the cold season, with no change of food which would be consistent with the alteration of temperature. The symptoms are known as swelling of the face, a scarlet eruption appearing on the skin in patches, sickness at the stomach, fits of heat, lassitude, vomiting, depressed pulse, and difficult breathing.

REMEDY.—Prevention consists in giving your children less stimulating food, and less in quantity of every kind, during the forty days of transitional weather at the close of winter and during the early spring months. From the last of February to the middle of April, you cannot be too watchful of your children's dietetic habits. Fresh fish may be used occasionally. Sparingly of milk, butter, and gravies. No meat of any kind; not a mouthful. But plenty of fruit, and bread, and simple puddings. No pies or greasy nuts.

If the disease is upon your child, with threatenings of malignant sore throat, do not delay the administration of warm water injections, two or three times the first day. Suspend every description of eating, and give the little patient very prompt attentions, calculated to inspire quiet and confidence. Bathe the legs and arms in water, hot with mustard, after which wrap them in soft flannel, and keep them very warm. This should be done at least once every day; twice, if the symptoms are violent. This will gradually overcome the throat irritation, and establish a balance of energy in the vitals. If the throat is violently threatened with diptheria or malignant symptoms, apply thin beefsteak compresses about the neck. Change as often and as quick as the piece becomes dry under the influence of the fever. This treatment, in connection with the usual homeopathic remedies, will scarcely fail to save the patient. Always magnetize the bowels, and hold and rub the feet and hands of the sufferer, after administering the water injections. Keep old-fashioned drug doctors out of the house; ditto all ministers of old school theology.

Origin and Philosophy of Fevers.

A positive condition of the blood is a perfect panoply against *fevers* of every shade and name. Fever is taken by the breath. The lungs absorb the contaminating effluvium, and the blood is forthwith poisoned. A cold negative state supervenes, and the vital heat rushes out upon the surface.

M. Roy has found that the fever which is so prevalent in Algeria is due to the fact that in the region of volcanic and primitive rocks, the clay contains phosphorus, and this, acted on by fogs and dews, which contain ammonia, diffuses its noxious qualities in the atmosphere, and occasions fever. By way of testing this theory, he created an artificial atmosphere of this sort, and, on breathing it, found he had all the symptoms of the African fever.

Continued and Exhaustive Fever.

The signs of a low, feverish, typhoidal state of the system, are: Sickness and sinkings at the stomach, heat and dryness of the skin, headache frequently, pain in the bones, weariness in the limbs, tongue dry in the morning, and white-coated, faintness and trembling while walking, with a costive state of the bowels, or a little relaxation.

REMEDY.—For this condition we know of nothing that will compare with a daily vapor bath and magnetic operations over the entire body. It is a mistake to eat anything after dinner. For dinner you should eat a pint or so of porridge, made of Indian meal and Graham flour, or rice, nearly equal parts, with water well boiled. Milk, sugar, salt meat, fish, cheese, cakes, pies, puddings, &c., are not allowable. You must work to put power into your vital system, otherwise the fever cannot be resisted; therefore, as the nausea, and faintness, and trembling subside, let your own appetite and judgment select the most agreeable diet. Drink a tea (half a pint) of white or pleurisy-root about every other day. Be very hopeful; the day of deliverance will soon dawn.

Diseases of the Eye.

We will just now write only of that subtile description of eye weakness, which is very general. Eyes so affected have a fixed debility in the posterior membranes and ciliary arteries. The entire nervous tunic in the back portions of the orbs is aqueous and feebly conditioned. The ciliary processes, both nervous and vascular, are de-magnetized and deficient in power. The brown and black matter of the eye—the *pigmentum nigrum* —is manufactured by the joint action of the ciliary arteries and their accompanying nerves. This material is also imperfectly formed and tardily secreted, and thus the eyes are unable to reflect or absorb the light promptly, which fact will account both for concealed inflammation and the weakness felt when reading. We find no disease in the ligaments, none in the choroid membrane (although so intimate with the ciliary mechanism,) none in the optic nerve, and the organism generally is in good condition.

CAUSES AND SYMPTOMS.—It will not comport with the objects of this department to enter upon lengthy details. Pages must be condensed and packed into short, perhaps ungraceful, paragraphs, and shorter still in brief sentences. Scientific knowledge and practical experience, in this day and age of the world, are tending to comprehensive conclusions, put into synthetical forms of expression. People, consequently, are called upon to read, to analyze, and to reason for themselves, on every imaginable subject. In one word, then, we affirm the cause of generally weak eyes to be *a loss of the necessary magnetic vitality from the ciliary processes.* The first cause may be connected with too steadily gazing at the sun, perhaps to observe an eclipse, but whatever the producing cause, the effects are as described, a hidden weakness and occasional inflammation, which no dieting, no virtue in habits, can ever perfectly remove.

THE REMEDY.—The diseased posterior portions of the eye cannot be reached by washes and internal treatment. The only sure and direct method is magnetic. The necessary equilibrium must be restored by means of the human hand, in conjunction with the electro-magnetic battery, unless some Indian spirit will inspire a medium to prosecute the treatment. Apply the negative pole alternately behind the ears; the positive pole in a bowl of water, in which the upper part of the patient's face should be immersed. This process for twenty minutes every night.

In cases where economy is necessary, we would substitute for the battery, after a thorough manipulation of the face and neck, and forehead, a bandage behind the ears loaded with a poultice composed of *white of an egg, a bit of unslaked lime powdered, and sprinkled with the black oxide of manganese.* It is wise, also, in severe periods of inflammation within the lids of the eye, to moisten the bandage with diluted *sulphuric acid.* A fresh preparation of this kind should be made every day, and more frequently in urgent cases. Let it be observed, once for all, that this semi-magnetic remedy for eye-weaknesses is effectual only in connection with human manipulation. If a healing medium can be obtained, so much more rapid and pleasurable will be the restoration. Any imperfect chylification of food is a clog in the wheels of success. Eat what you need; not what you want, unless healthy. *Only the good are free!* "The way of the transgressor is hard."

Chronic Sore and Weak Eyes.

Follow the hygiene and general laws of health prescribed for the cure of eruptive diseases. Put cold water on the back of your neck and behind your ears every morning. Do this thoroughly first; then wash the face as usual. Manipulations on the temples are very useful, especially after going to bed for the night. In addition to the foregoing, but always after removing all dampness occasioned by the morning ablution, bathe and saturate the eyes with four table-spoonfuls of sweet milk, in which you should pour half a tea-spoonful of laudanum. This treatment will also serve for weak and watery eyes, with this exception—tepid water should be substituted for the sweet milk.

Blurring or Dimness of Sight.

In all cases of eye-disease the patient must not be permitted to use salted food, nor saleratus in any form with daily nourishment, as both are irritants of the poisonous kind to every delicate membrane, and to the liver. If the taste calls for "salt," it is wisest to place a little on the tongue, simply to allay the wish for it, but do not swallow much of the salted or alkalined saliva. Many eye-diseases, which began with inflammation, may be cured by anti-salt and anti-saleratus habits at the table. Use no salted bread, nor butter, nor salted meat of any kind, while suffering with throat or eye disturbances. Fresh fish will not be palatable, and should not be used under this treatment; but eggs, and cooked fruit and vegetables, may be advantageously used. With these rules employ human magnetism from head to feet. A diseased human body should be rubbed and dressed down by the hand, somewhat as favorite horses are manipulated by their keepers. Many persons would be extremely grateful for treatment as kind and tender as that bestowed upon the expensive trotter in the stable. Make a tea of blood-root, (*sanguinaria canadensis*,) and bathe the eyes with it twice or thrice a day.

Ophthalmia, or Inflammation of the Eye.

This painful inflammation of the eye is usually produced by a cold wind, dust, or some external irritation. It generally begins with the appearance of a net-work of blood-vessels on some part of the conjunctiva of the eye-ball or eyelids. The eyelids become swollen and tender, and the redness soon covers the whole conjunctiva; there is increased discharge of tears, intolerance of light. There is more or less of constant pain, and a sensation as if particles of fine sand had insinuated themselves under the eyelid, accompanied by a great heat and pricking pain. A glutinous matter is now secreted, especially in the night, which causes the eyelids to stick very firmly together.

REMEDY.—It is no unusual thing for the disease to commence in one eye, and in a day or two to seize the other. Mild cases of conjunctiva ophthalmitis generally run their course in a few days, and cease spontaneously, or are removed by a purgative, abstaining from the usual diet, and remaining in a dark place. The best local applications are: A warm or filtered decoction of poppy leaves, or one fluid drachm of the tinctura opii in eight fluid ounces of distilled water. The eye should be well cleansed from the glutinous matter with warm milk and water.

31*

Paralysis, and Total Blindness.

Blindness is sometimes sympathetic—a result, through the involuntary system, of physical suffering—involving the middle stratum of the brain. That the optic nerve is paralyzed, to some extent, no one acquainted with the structure of the eye can doubt; but this fact in physiology does not preclude the effectual application of restorative influences.

REMEDY.—Hold the head down on rising every morning, so that an attendant can pour from a hight of three or four feet, a quart or more of cold water, steadily on the back of the neck, behind the ears, and on the lower portions of the brain. This practice should be continued until an eruption or boils come out on the part so treated, or until the eyes begin to take on more sensitiveness to the sunlight, when it is best to commence the periodic application of hand-magnetism by manipulation to the neck, temples, eyes, and lower brain, including the face and shoulders.

Obliquity of Vision.

The sight of organs may be good; and yet it is possible to have obliquity of vision. Only a partial relief can be rendered by artificial means. Archery is a fine discipline for the vision. Ninepins and billiards very injurious. It is our impression that a certain kind of spectacles, usually worn by persons in the first stages of old age or weak sight, will assist greatly in correcting the defect. Hand-magnetism would be useful at times. Avoid night reading. And always look *squarely* at any object. To do this well, employ your Will.

Spots, Webs, &c., before the Eyes.

The cause of these symptoms is traceable to a dormant condition of the liver, affecting the spleen and kidneys.

REMEDY.—The right side must be thoroughly pressed and

kneaded—sometimes pounded every forenoon, so that the bowels and kidneys will act promptly and freely. Sluggishness of the intestines will bring "spots" before the eyes, and occasional blindness. The sight depends almost exclusively upon the health of the nervous system.

White Spots on the Eyelids.

REMEDY.—Treat them as you would a mole or wart—viz.: Take a hair and tie it tight around the base of the excrescence. By cutting off the capillary circulation, and stopping the flow of the nerve-life, the blemishes become foreign bodies, decay, and soon drop off. Perhaps it will be necessary to keep the ligature on for many days. If you cannot tie a thread at the base of the mole or wart, the true substitute is to use a needle to draw a thread through the excrescence, leaving the two ends about an inch long. Occasionally pull the thread back and forth, and the contents of the spot, wart, or mole, will soon run out.

The Eye and the Mind.

The mind (says a writer,) like the eye, has its adjustment to near and remote objects. A watchmaker can find the broken tooth in a ship's chronometer quicker than the captain, and the captain will detect a sail in the distance long before the artisan can see it. Physiologists and metaphysicians look at the same object with different focal adjustments.

Wash for Sympathetic Sore Eyes.

Take a large pinch of bayberry bark (*Myrica Cerifera,*)—not "barberry," remember—make a strong tea of it; when cold, wash the eyelids with it, both at night and first thing in the morning.

If there be *granulations on the lids*, as an effect of inflammation, take equal parts of table-salt and citric acid, say half a tea-spoonful of each, and mix them in the white of an egg. Bandage your eyes with this every night. Never use the same mixture twice. Make it fresh just before applying.

If there be much *itching of the lids*, use the water of turpentine, which may be obtained by pouring rain-water on crude gum taken from the pine tree; six ounces of gum or "pitch" to half pint of warm water; let this tincture stand two or three days, then apply several times during twenty-four hours. Persons with sore eyes must *not* use salted food.

If there be temporary *loss of vision*, it will then be necessary to arouse the nerve-energy of the best eye by the use of hand-magnetism. Pouring cold water on the neck every morning is good, as a tonic, but is not likely to do what magnetism can. Do not delay the application of the proper means for your recovery.

There is a sort of *rainbow* in every eye—the colored membranes around the pupil—called the "iris." Watch it closely, when the light is in proper angle with the sight, and you will see a rainbow. What does it "promise"?

Nutritious Food for the Sick.

The *Eclectic Journal* says such food may consist of vegetable soup prepared by boiling pieces of turnips, potatoes, rice, or cracked wheat or rye, and adding small quantities of fresh dried beef, allowing the patient to take a wine-glass full, three, four, or five times a day, if required. Dropped egg, egg toast, bread toast, bread coffee, rye and wheat gruel, are also admissible, and many other kinds of farinaceous gruels and broths.

Pleasant Drink for the Sick.

We prescribe for the sick, everywhere, the following delicious beverage: To one and a half pounds of white sugar, two table-spoonfuls of brewers' yeast, put in one gallon of water. Let it stand in a warm place till it ferments. Then strain and bottle it. Keep it in a cold place. It should not be used until the day after it is put away in well-corked bottles.

Poultices for Wounds and Injuries.

These may be made of slippery elm, flax-seed, bread and milk, etc. A very convenient flax-seed poultice is made by taking flax-seed malt, and adding boiling water, until it assumes the consistency of a poultice. Slippery elm poultice may be made in the same manner. Bread and milk poultice may be made, by taking equal parts of each, and boiling them until they are in the form of mush. Potato and carrot poultices are made by boiling these articles, and mashing them fine. Poultices should be spread on small pieces of cloth, laid on to the part, and protected by a small piece of oiled silk.

Water Dressings for Wounds.

Many surgeons now-a-days are in the habit of applying water-dressings exclusively to wounds. These are made by taking several folds of linen or cotton, wrung out of cold or tepid water, and applying them to the part. Change as often as necessary to keep the parts moist. They should be covered with oiled silk, in the same manner as poultices.

Rapid Cure for Burns.

The *Medical Gazette* of France says, that, by an accident, charcoal has been discovered to be a cure for burns. By laying a piece of cold charcoal upon a burn, the pain subsides immediately. By leaving the charcoal on one hour, the wound is healed, as has been demonstrated on several occasions. The most painful, but most sure and positive, remedy for a bad burn or scald, is the *tincture of lobelia* used as strong as the patient can bear it.

Treatment for Poisoned Wounds.

A disciple of Hahnemann says: The application of dry heat at a distance is an excellent remedy against the bites of venomous serpents, mad dogs, etc. This remedy is strongly recommended by Dr. Hering. A red-hot iron, incandescent coal, &c., ought to be placed as near the wound as possible; without, however, causing too sharp a pain; this should be continued until the patient begins to shiver and stretch himself. If this should take place a few minutes after the heat is first applied, the application may be continued for an hour, if the patient is able to bear it, or until the effects of the poison disappear.

Excellent Mixture for Burns.

According to good medical authority, two parts of collodion and one of castor-oil make an excellent mixture for burns. The mixture should be spread on with a camel's-hair pencil. It forms a covering that protects the parts from the air and other irritants, and is perfectly painless. It may be allowed to remain until suppuration begins, when a poultice of light bread and water will remove it. The healthy granulating surface may be dressed with simple means, and it readily heals. Should cracks occur in this artificial cuticle, a little more of the article will close them up. Two coats, put on the first and second day, will answer the purpose.

Remedy against Poisoning by Phosphorus.

Poisoning by phosphorus has become so frequent, in consequence of the universal introduction of chemical matches, that it is highly important to make every one acquainted with the best means of counteracting the effects of that substance, in order that speedy relief may be afforded to those who may have had the misfortune to take some. The *Medizinisch-Chirurgische Monatschrift*, a German medical paper, proposes calcined magnesia as the best remedy for the purpose, stated to have been largely experimented on by Doctors Antonelli and Borsarelli. In cases of poisoning by phosphorus, or by any other substance containing that metalloid, the administration of fatty substances should be avoided; because, far from attenuating the effects of the poison, they increase its energy and facilitate its diffusion. Calcined magnesia should be administered in large quantities, suspended in water that has been exposed to ebullition. In cases of dysuria occasioned by phosphorus, the best remedy is acetate of potash. All mucilaginous beverages administered to the patient should be prepared with water that has boiled, in order that they may contain as little atmospheric air as possible.

Antidote for Strychnine.

Tannin, which retains nicotine, unites also with nearly all the other alkaloids. Kursak recommends it as an antidote for strychnine, in doses of sixty grains of nut-gall to one of alkaloid. Nearly any astringent vegetable substance which happens to be at hand, may replace the tannin, such as the glands or bark of the oak and of the chestnut. The use of acids and alcohol should be avoided, as these agents dissolve the poison.

Results of a Useful Emetic.

A small child having got a cherry-stone lodged within its nose, and the efforts of a regular practitioner having failed, the services of the village barber, who likewise practiced the heal ing art, were called in. He administered a powerful emetic, and at the moment when vomiting was about to commence, clapped a handkerchief tightly over the child's mouth. Either from the violent expulsion of the contents of the stomach against the nares, or from the impulse given by the air expired, the stone dropped on the floor.

Swallowing Lizards, Frogs, and Toads.

Bertholin, the learned Swedish doctor, relates strange anecdotes of lizards, toads, and frogs; stating that a woman thirty years of age, being thirsty, drank plentifully of water at a pond. At the end of a few months, she experienced singular movements in her stomach, as if something was crawling up and down; and, alarmed by the sensation, consulted a medical man, who prescribed a dose of orvietan in a decoction of fumitory. Shortly afterwards, the irritation of the stomach increasing, she vomited three toads and two young lizards, after which she became more at ease. In the spring following, however, her irritation at the stomach was renewed, and aloes and bezoar being administered, she vomited three female frogs, followed the next day by their numerous progeny. In the month of January following, she vomited five more living frogs, and in the course of seven years ejected as many as eighty. Doctor Bertholin protests that he heard them croak in her stomach!

Poisoned by Saltpeter.

This poison produces violent symptoms, and acts like many other poisons, with fatal promptness. TREATMENT.—Any emetics which excite and irritate the stomach, are very dangerous. Vomiting, however, should be immediately induced by large and constant draughts of light mucilaginous drinks—flaxseed tea, gum water, skim milk, sweet oil, barley water, slippery elm tea, either of which may be used to induce vomiting, and thus save the patient's life. After the removal of any poison, it is best to eat and drink only small quantities of the simplest preparations.

Poisoned by Ash-Leachings.

Vinegar and ley, by uniting chemically, form the acetate of potash. Oil and ley, united, form soap. Therefore, should a person, by mistake, drink ash-leachings, the natural antidote, to save life, would be a large draught of either sweet oil or vinegar, immediately administered. During the twenty-four hours succeeding, give the patient an abundance of flax-seed tea; almost no food of any kind; and lay cold compresses on the stomach and abdomen.

Bee or Serpent Poison.

For the bite of any venomous creature, reptile, spider, bee, or locust: Mix gunpowder and sweet-oil together (just enough oil or lard to make the grains of the powder adhere,) and apply immediately; do not fear; the result will be favorable. Change the poultice several times during the day.

Persons Stricken by Lightning.

Never use the magnetic current upon persons stricken down by lightning. Open their clothing at once, giving free scope to lungs and blood, and drench them with frequent buckets of cold water. Try artificial respiration also—*i. e.*, breathe into the patient's mouth, pressing the breast immediately after inflating the lungs—and apply vigorous friction to the hands and feet.

Cure for Foul Air.

When an unpleasant odor (tobacco excepted,) is found in the sick room or elsewhere, if a few grains of coffee are scorched or roasted (not burned) in such places, the disagreeable odor will soon depart. Casks, or other vessels, also clothing, can likewise be cleansed in this manner. Those who have not tried it, will be surprised to witness its *quick* and *positive* effect.

Removal of Superfluous Hair.

It sometimes happens that the conditions for the production of hair exist in parts of the face or body where Mother Nature did not design to fix them. In such cases, the tuft of extra growth is many times a source of annoyance. It should be understood that each hair, in growing through the skin, carries with it a transparent sheath. Each hair, in fact, consists of several hairs; among them, there is a passage for the growing fluid to flow out and in. The true way to obliterate hair, therefore, is: To eradicate the *minute canals* which convey the nutrient liquids to the visible stems. This can hardly be done without injury to the skin. The simplest remedy within our knowledge, in this country, is the active principle of *Colchicum*, (get the alkaloid extract); moisten with weak vinegar, apply it to the parts after shaving the hair down, and cover the medicine with court-plaster. Renew the remedy a few times; keeping the air from the parts under treatment. You should put great faith in Time—

> "Time, the beautifier of the dead,
> Adorner of the ruin, comforter
> And only healer when the heart hath bled—
> Time! the corrector where our judgments err,
> The test of truth, love—sole philosopher,
> For all the rest are sophists."

Restoration of the Hair.

Tincture of cantharides, one drachm; orange-flower water, eight ounces; sugar of lead, half a drachm; sulphur, three quarters of an ounce. Shake this mixture thoroughly on using it. We can assure every lady, unless her bodily state is very low and feverish, that the foregoing is the surest remedy for diseased or itching scalp, falling hair, and premature turning of color. The hair must be first combed carefully with a leaden comb; then saturated with the above restorative at bed time; next morning the head may be washed with a little soap and much water. Patients troubled with headache after dressing long hair, should remove the extra length with the scissors.

The premature loss of hair is sometimes attributable to an alkaline deposit, which can be removed by frequently washing the scalp with weak sulphuric acid. Only ten drops to half a pint of rain-water.

A Genuine Hair Tonic.

The following formula, having been fully tested by our "Angel of the House," and pronounced "good" by the authority of her experience, is hereby recommended to the million: Take one ounce sugar of lead; one ounce lac. sulphur; half ounce oil of bergamot; one gill of bay rum; one tea-spoonful of salt. First dissolve the oil of bergamot in half a gill of alcohol; then add all the ingredients to one gallon of rain-water, bottle it up tight, and it will be ready for use. Use it once a day, rubbing the scalp thoroughly.

How to Wash your Mouth.

Cleanse all parts of your tongue and teeth, punctually before breakfast and directly after supper. This simple act of devotion will silently sweeten your whole body. You should remember that man's dental organism is not subject to climatic influences. If your Causality be developed, we advise you, first of all, to look for the causes of defective teeth in the nervous

American's ill-ventilated lungs; next, in the gallons of malarious ethers that emanate from ill-digested food in the diseased duodenum. It will be seen that bad breathing is a great cause of bad teeth.

Origin of Bitter and Sweet.

Dr. W. Herschel has discovered that the mixing of nitrate of silver (lunar caustic) with hypo-sulphate of soda, both remarkably bitter substances, produces the sweetest substance known; a proof how much we are in the dark as to the manner in which things affect our organ of taste. So, bitter and sweet, as well as sour, appear not to be an essential quality in the matter itself, but to depend upon the proportion of the mixtures which compose it.

Thin Shoes and Wet Feet.

It is useless to amplify and moralize upon this fertile source of female suffering. Thousands pass from girlhood into the grave—but many times not before they have injured the world by the bestowal of one sickly child—solely in consequence of habitually wearing *soleless* shoes. REMEDY.—Get wisdom, take exercise, breathe plentifully of pure air, add a pair of easy, thick-soled shoes, and improve the style of your dress.

Onions and Cider as Medicines.

The popular roots (called onions,) are filled with medical properties. The magnetic power of the compass-needle will be entirely changed or destroyed by the touch of onion-juice. Human instinct first used them in poultices.

We do not deem apple-wine or cider good as a beverage; at times it is both anti-bilious and cathartic; in general, it is the reverse in effect. See preceding pages for our doctrines regarding Foods and Drinks.

Origin of the Human Spine.

The spinal cord is an extension of the mammalial world into the human organization. Spines pre-existed in the organic sphere for ages; the brain was an after development. Say, rather, that the brain is composed of a multitude of spinal cords.

A Palm Blister as a Remedy for Disease.

We proclaim another mode of treating disease (says the *Scientific American,*) — a treatment that casts homeopathy, hydropathy, steam doctoring, the movement cure, and the science of therapeutics itself, entirely into the shade. It is well known that all these systems, though they make a loud noise in the world, really accomplish very little; nearly all patients who recover under the treatment of physicians of any school, would have recovered without the aid of the physician, and it is very seldom indeed that fatal diseases are diverted from their course by putting drugs into the stomach. But our system is effectual; it will cure many of the worst diseases to which mankind are subject, and it will prevent them all. It is as simple as it is powerful; it is nothing more than raising a blister in the palm of the hand. The blister must not be raised by cantharides or other poisonous irritants, but must be produced by friction, accompanied with an alternate contraction and extension of the muscles. If the operation acts as a sudorific, inducing a sensible perspiration between the clavicles and above the eyebrows, it is all the more efficacious. Almost any solid substance may be employed for administering the friction, though it has been discovered that the best substance for the purpose is the handle of some tool, such as a hammer, saw, or plane; the very best of all being the handle of a plow or hoe.

This treatment produces the good effects of all the articles in the whole materia medica, and with more power and certainty than they. For instance, it is a more powerful opiate than opium, and, while the sleep induced by narcotics is succeeded by nausea and debility, that resulting from this treatment is wholly refreshing and invigorating, and is followed by a peculiarly healthful and buoyant exhilaration.

As a tonic it is more beneficial than bark or iron, not only strengthening the muscles, but actually enlarging their volume.

To give an appetite, it is better than any dinner pill. If the epicure, who sits down to his table with indifference, and forces a few mouthfuls of his dainty viands into his stomach, where they give him great distress, will adopt this treatment, he will come to the table with a keen desire that will give a relish to the plainest food; and digestion waits upon an appetite thus produced.

It is a better remedy for incipient consumption than cod liver oil, and is a sovereign cure for dyspepsia, jaundice, liver complaint, and a long train of chronic diseases.

It will not only remove bodily ills, but is the best of all medicines for a mind diseased. If a man who is suffering from hypochondriasis, who feels that the burdens of life are greater than he can bear, and who sees the clouds of despair settling over his future, will take hold of a shovel-handle and raise a blister in the palm of his hand, he will be surprised to see how the troubles that have oppressed him are brushed away, and the future before him is brightened. New beauties will come upon the face of Nature, and new joys and hopes will spring up in his heart. This is the true elixir of life.

While other modes of treatment are expensive, this not only costs absolutely nothing, but it is a source of revenue to the patient.

It removes not only sickness and despondency, but poverty also. It is a remedy for all the ills that flesh is heir to.

Though this system is the best of any for the cure of complaints, its great superiority is as a prophylactic. If properly administered to a healthy subject, it will prevent all disease. The next neighbor to the writer of this died, at ninety-four, of old age. A few days before his death, in conversation with him, we asked him if he had ever consulted a physician. He replied that he never had.

"Were you ever sick?"

"No."

"Not a day?"

"No."

"Not an hour?"

"No."

"You were never sick in your life?"

"No."

This man knew nothing of physiology; he had never practiced any system of dieting; but every day, Sundays and all, for more than eighty years, and generally (excepting Sundays) through the whole day, from before sunrise till after sunset, *he had applied friction to the palms of his hands.*

An Antidote for Putrefaction.

In all malignant diseases there is a rapid tendency toward putrescency, which may be obviated by the use of several antiseptic medicines, either in the form of solids or as vapors. Among the best may be mentioned acetic acid, sulphate of iron, corrosive sublimate, sulphate of copper, creosote, chlorine, acetate of iron, chloride of lime, soda, alcohol, and quinine. In houses infected by the poison-vapor of small-pox, or by other epidemic and malignant diseases, we recommend the constant burning of *a chlorine lamp*. The vapor will diffuse itself through all parts, and prevent the general tendency to putrefaction.

The Medical Uses of Water.

We would remark, for the good of all invalids, that the *temperature* of water is the positive essential. Cold water, ranging from 60 down to 32 degrees, is charged with *electricity;* while hot water, or vapor, from 100 to 140 degrees, is *magnetically* charged; and much of the medicinal effect of this element must, of necessity, be in accordance with its temperature. The scale of temperatures, according to the latest publication, is thus given: Cold, 32 to 60 degrees; cool, 60 to 70; temperate, 70 to 80; tepid, 80 to 85; warm, 85 to 100; hot, 100 to 120; vapor, 110 to 140.

The father of the hydropathic system, Priesnitz, was strictly a "cold water" physician. But his success with disease was not very remarkable; and so, we observe, many of his followers have enlarged their views and have improved his system. Every hydropathic institution in America, of any consequence, has added the *movement cure* and some *magnetic treatment* to their improved applications of water. Their patients, as a general rule, leave with better complexions and higher hopes. We have great faith in water as a preventive of disease, and think it should be more used.

Nervous and Convulsive Diseases.

All such diseases can be cured if the patient will bestow particular attention upon the application of Food, Water, Air, and Magnetism. The application of zinc and copper plates to the feet, after bathing the latter in cold water, will be of great service at night. *Time*, and not medicine, is the best remedy for such complaints, especially when the individual is obeying the laws of Nature, and striving to repel, *by Will of mind*, the malady from the system. Let *spirit* arise superior to the visible form; keep the latter continually negative and the former positive. A strong-minded man is seldom diseased; he is determined not to be subdued by various afflictions, for he is superior to them, hence *Wills* and repels them away. So should our patients do, assisted by the spiritual influence of some congenial person. The principal food should be solid and nourishing. Watery substances for diet tend to weaken the blood, solids, and muscles. Exercise should be moderate, and *early in the morning.* Patients should be very careful to avoid all excesses. Study, or passionate exercises of the mind, must be abandoned, and the unequal action of one class of organs of the body or mind must not be permitted. Sleep an hour between breakfast and dinner.

How to Retain a Good Face.

A correspondent of the *Home Journal* has some good ideas on the importance of mental activity in retaining a good face. He says: "We were speaking of handsome men the other evening, and I was wondering why K. had so lost the beauty for which five years ago he was so famous. 'Oh, it's because he never did anything,' said B.; 'he never worked, thought, or suffered. You must have the mind chiseling away at the features if you want handsome middle-aged men.' Since hearing that remark, I have been on the watch to see if it is generally true, and it is. A handsome man, who does nothing but eat and drink, grows flabby, and the fine lines of his features are lost; but the hard thinker has an admirable sculptor at work, keeping his fine lines in repair, and constantly going over his face to improve the original design."

Cholera Morbus and Cholera Flatulenta.

All severe forms of cholera begin with symptoms of flatulency; with a sense of oppression, soreness, pain, and distension, in the stomach and throughout the bowels; all which is quickly succeeded by great depression of Will and spirits, and by severe vomiting and purging, and a clammy sweat all over the body; sometimes there is much difficulty in breathing, hiccough, irregular pulse, convulsions, cramping of the cords in the legs, and a coldness of the skin, while the patient persists in calling for cooling drinks, thinking that he is burning up with great heat.

REMEDY.—Do not fear anything, but make up your mind not to die in that unbecoming manner. The cause of all your suffering is to be traced to some imprudences of your own. Stomach and liver are resisting the evils of your recent violations. Put mustard and onion poultices on the feet, extending up the legs; also poultice the hands, wrists, and the stomach and bowels; or use plentifully of hot-water cloths over the stomach and intestines. Eat nothing. Drink weak lemonade and flax-seed tea, without sugar; or lime-water, in order to stop the vomiting and quiet the bowels. Abstain from solid food for many days. If the cholera should result in a dysentery, then treat the patient according to directions for that disease.

Treatment for Typhoid Fever.

September is a fearful month to persons who are predisposed to a low grade of typhoid fever. The summer time, with its peculiar influences upon the brain and blood, in passing away, is certain to leave a "strait" for mankind, filled with malarious vapors and bilious fevers. There are three predisposing and producing causes of eccentrical epidemics, viz.:

1st. The miasmatical character of the location, and the conspiring effluvia of its environs.

2d. The position of the location with reference to peculiar longitudinal *magnetic currents* of the earth.

3d. The situation of the street, or city, or other location, as determining the degree of light received from the sun.

In different places a low form of typhoid debilities prevail—with a slow, consuming fever—but without producing the ulcers in the bowels, and lesions, which are the well-known characteristics of the regular typhoid.

REMEDY.—Stop all food as soon as your head begins to ache. Drink nothing but cold flax-seed tea, or the tea of slippery elm, with a few drops of lemon-juice. Magnetism is a great remedy if applied vigorously to the extremities in the early stages of the disease. The Water Cure system is the best for the symptoms of this fever. Nothing can be more remedial, or more grateful to a hot skin, than a wet-sheet pack once or twice a day. But we think there is no danger from this fever, if the patient will abstain from food on the first noticeable symptom, and gradually return to his customary diet after his appetite is fully restored. Quinine and calomel are popular but dangerous medicines. They leave the patient with some other disease.

The Physical and Spiritual Man.

Man is a UNIT. It is not true that he has a *body* to be cured of disease separate from his *mind;* nor is it true that man has a *spirit,* a *soul,* a *heart,* to be cured of sin-diseases separate from his body. The physical and spiritual organizations of man are, in this rudimental or caterpillar state of existence, "one and inseparable!" If clergymen suppose (as many of them most conscientiously do,) that the moral and religious sentiments and qualifications of the human *soul* can be touched and unfolded into practical exercise merely by preaching sacred principles to it, then we are impressed to undeceive them. And if physicians believe (as many of them profess to,) that the human *body* can be cured of its endlessly modified afflictions *merely* by administering scientific preparations of mineral and vegetable substances, then we are also impressed to undeceive them. It is absolutely impossible to develop realizations of heaven in the soul when that soul is not attuned to perfect harmony. From various causes, which have been fully explained, the animating essence of the human body is thrown or pressed into different degrees of discord; and the relation between this essence and every organ, nerve, and muscle, is so inconceivably and inexpressibly intimate, that the body becomes the *day-book* and *ledger* in which are recorded the most *trivial* as well as the most *complicated* of disturbances which the spirit is made to experience. The enlightened mind, therefore, cannot but perceive that any unsettled accounts between the human soul and external nature will act as positive obstructions to the development and exercise of pure religious principles. But how surprisingly unphilosophical are the teachers of the present generation! How unphilosophical and useless to preach and complain that the human heart is slow to perceive truth, is inclined to evil and sin, that it resists the saving and momentous truths of heaven, when, from some cause, the soul—the entire individual—is suffering from the melancholy effects of dyspepsia, or constipation, or from other constitutional inharmonies!

The Physiology of Courage.

Emerson, in his latest work, gives the physiology of courage: Courage—the old physicians taught (and their meaning holds good, if their physiology is a little mythical)—courage, or the degree of life, is as the degree of circulation of blood in the arteries. "During passion, anger, fury, trials of strength, wrestling, fighting, a large amount of blood is collected in the arteries, the maintenance of bodily strength requiring it, and but little is sent into the veins. This condition is constant with intrepid persons." Where the arteries hold their blood, is courage and adventure possible. Where they pour it unrestrained into the veins, the spirit is low and feeble. For performance of great mark, it needs extraordinary health. . . . . . . There is no chance in results. With adults, as with children, one class enter cordially into the game, and whirl with the whirling world; the others have cold hands, and remain bystanders, or are only dragged in by the humor and vivacity of those who can carry a dead weight. The first wealth is health. Sickness is poor-spirited, and cannot serve any one : it must husband its resources to live. But health or fullness answers its own ends, and has to spare, runs over, and inundates the neighborhoods and creeks of other men's necessities.

Tea, Coffee, Cocoa, and Chocolate.

Coffee, or tea, or chocolate, when very strong and very hot, are very injurious. It is not the tea, nor the coffee, that is injurious to the constitution, but it is their strength, their too great heat, and their excessive use. Cold coffee is sometimes a pleasant and highly valuable tonic. Tea is not very injurious, and weak cocoa and chocolate are both important beverages in some lingering and nervous complaints.

Sweet Oil as a Remedy for Poison.

"It is now over twenty years," says a dairyman, "since I learned that sweet oil would cure the bite of a rattlesnake, not knowing that it would cure any other poison. Practice, observation, and experience, have taught me that it would cure poison of any kind, both in man and beast. I think no farmer should be without a bottle in his house. The patient must take a spoonful internally and bathe the wound for a cure. To cure a horse it requires eight times as much as a man. Here let me say, one of the most extreme cases of snake bites in this neighborhood, eleven years ago this summer, when the case had been over thirty hours standing, and the patient given up to die by his physician, I heard of it, carried the oil, gave him one spoonful, which created a cure. It is an antidote for arsenic and strychnine. It will cure bloat in cattle, caused by eating too freely of fresh clover ; it will cure the sting of bees, spiders, or any other insects ; and it will cure persons who have been poisoned by low running vines, growing in meadows, called ivy."

[The reader is prepared to believe that our confidence in the sovereign virtues of olive oil is strong, from the frequency with which it is prescribed in this volume. But we do not indorse the merits and uses of sweet oil to the above extent. So much reliance upon its power to antidote every kind of poison, would be attended with great and fatal disappointment.]

Remedies for Fever and Ague.

Intermittent Fever, or Fever and Ague, as we have before written, illustrates the origin and philosophy of all human diseases. It will be seen that the temperature of the body is thrown into a positive state by certain electrical conditions of the atmosphere, and into a negative state by conditions which are exactly opposite. The negative condition is cold, and the positive warm. In other words, the positive state is the *feverish* condition, and the negative state the condition of *chill*. Fever and chills in the atmosphere, therefore, develop and strengthen *fever and chills* in the human system. This atmospherical condition can and does exist a long time, in some seasons and countries, before the resisting power of the human body is overcome. But the physical structure, like the spiritual structure, is ever subject to the influence of surrounding conditions and circumstances; and the power which these conditions and circumstances possess, is not only sufficient finally to overcome the resisting power of the body, but they first throw the mind itself out of health, harmony, and due proportions. The abounding dampness and electricity (which are negative,) *contract* the cuticle glands, relative membranes, and serous surfaces of the organization, and thus are *repelled* the spiritual forces and fluids which reside in and circulate through them when the healthy temperature and condition exist. The consequence of long-continued disturbances of this kind, is a chill, which soon reacts into a fever; and thus is established the intermittent complaint. The fever is occasioned by a partial return of the forces and fluids to their appropriate places on the external surfaces.

The difference there is between intermittent fever and other spasmodic complaints, consists in this: In Fever and Ague there occurs an incessant succession of spasmodic motions during the whole paroxysm; while in the other affections these motions are more concentrated and conspicuous; but in every spasmodic disease, the same muscles are affected in the same manner, and by the same primary causes, differing from chills and fever only in degrees of violence and frequency, according to which difference they have been branded with a Greek or Latin name by the medical profession. If an individual has once had chills and fever, he is liable to a recurrence of the disease at any time—especially whenever a heavy cold is taken, or the bodily temperature is changed. The disease is simple however, and its cure is correspondingly easy and natural.

If we have been enough fortunate to fully impress the reader's understanding with the true Philosophy of Disease, he will not need to be reminded in this place that it is the nerve-spirit (or force *within* the nerves,) which shakes and trembles in the cold or chilly stage of this disorder. No man's nerves would stir if his spirit was withdrawn from them. It is the dynamic life of the mind—the force, the energy, the power *within* the nerve—that is disturbed. Hence the spirit, and its Will, are the chief agents of cure. It is this fact, underlying the ten thousand "charms" practiced by superstitious "seventh sons" and credulous old ladies, by which many Fever and Ague patients have been instantly healed. But what will cure one in a few days, or hours, perhaps, would exert no remedial power upon persons of different organizations. The success or failure of psychological "charms" among the sick, is wholly referable to temperament—which law is strikingly illustrated in religious revival meetings, where, under the enchanting God-spell (*i. e.*, Gospel,) imparted from the pulpit orator, one person is straightway "converted and saved," while another remains cold, untouched, uninterested, and, therefore, unchanged.

REMEDY.—If a person has been long afflicted with this nerve-chill, with its accompanying headache, resultant fever, and ultimate prostration, it will be necessary for him to leave the country which brought the disturbance upon him. It is within the power of every person to *prevent* attacks of this disease, simply by keeping his appetite within bounds, discarding gravies, fat meat, butter, hot drinks, and newly-baked bread, and not working his strength down to a low point in the spring or autumn. But the "pound of cure" is most in demand, and that doctor is considered the "cleverest," and most "wholesome" to send for, whose doses are largest and most energetic in their operation. If an ignorant man pays fifty cents for a dose of medicine, he wants his "money's-worth" in *quantity* of the article to be swallowed. The "quality" is of little account in his estimation. Many ignorant, but good people, swallow *calomel* and *quinine,* as if it were within the power of such minerals and barks to heal their bodily infirmities.

Before prescribing for your Fever and Ague—or chills and fever—or intermittent disturbances—all the same thing under different names—we urge you to remember that your restoration will depend upon *the promptness and energy with which you exert your own Will.* All medicines are sometimes liable to fail. Nature's laws will never fail. They are the life of God. They cannot be changed, nor staid in their slow, calm, eternal round of operation. They tell us, infallibly, that MIND IS MASTER OF MATTER. So let it be!

When you feel the chill coming on, prepare your mind to resist it, arise to your feet, walk, or enter upon gymnastic exercises, and do everything to bring up the arterial circulation. Your coldness is owing to the blood principally occupying the veins, thus depriving the arteries of their customary magnetism and warmth, and producing a sensation of cold or chilliness all over the body. If you *wait* for the reaction, then you get an unnatural heat, which is the prostrating fever. Do not wait for such slow reaction.

The following is a sure remedy for breaking up the chill: Get one gill of best brandy, put in it a table-spoonful of fine salt, mix thoroughly, and take a wine-glass full on the first sensation of the ague. The influence of salt on the sympathetic and pneumogastric nerves is very surprising. It is well known that salt will counteract the action of brandy in the human stomach, so that a very drunken man may be perfectly sobered in less than an hour. Salt and water will stop a hemorrhage in the head, nose, or stomach; the same, tea-spoonful at a time, is good for stomach worms. If an aguish patient should try the above remedy, he will be astonished at the relief which it will bring to his shaking nerves. In some cases it may be

33*

necessary either to reduce the quantity, or take several doses, before the exact point of benefit is reached. The true way to cure this disease, is, to meet the chill with both your Will and your remedy, promptly and energetically.

Phthisic, or Asthma.

When a child is badly afflicted with the symptoms, give a tea-spoonful of pure linseed-oil in a table-spoonful of white brandy. In severe cases, add to the oil and brandy twenty to thirty drops of tincture lobelia, or less of ipecac. Also dip a piece of brown paper in a solution of saltpeter, let it dry, and then burn it so you can inhale the vapor.

If the lungs are weak, then, just before eating your first and second meals, take about twenty drops of pure olive oil. Take in your mouth, mix it with your saliva, and swallow slowly. Let all cases of *chronic dyspepsia* try this simple remedy for a number of weeks.

Pin Worms in Ano.

The quickest cure for "seat worms," so-called, is very strong salted water. First moisten the finger in the white of an egg, then immerse it in the salt water, and thus introduce the remedy up the anus. A few applications will relieve the most inveterate cases.

Catarrh and Difficult Breathing.

For chronic catarrh and difficult breathing—symptoms, profuse and fetid discharge from the nose, and occasional sore throat—take a tea-spoonful of sweet oil *early* in the morning and last thing at night. Also rub the breast and throat with the same every night and morning. Persist in this, and you will get well.

We admonish all this class of patients to be exceedingly cautious of sudden changes in bodily temperature. Bathe the feet in cold water before going to bed. Then give them a thorough coating of sweet oil ; next, draw on flannel socks, such as you do not wear daytime ; and lastly, take a tea-spoon half full of linseed-oil internally. Also a few drops of the same any time during the day, whenever your cough is dry and the pain troublesome

Tenderness at the Pit of the Stomach.

If your female friend has an inflammatory disease of the mucous membrane of the stomach, the " pit " will be tender, or sore to pressure, showing that the pneumogastric nerves are much disturbed. She must avoid "spoon victuals "—such as bread and milk, &c.—in short, everything that distends the stomach with gas. Slowly her health will improve, if she is careful to obey the laws of life ; in this we include regular kneading, and also manipulations over the stomach and bowels For painful menstruation, lay a light flax-seed poultice on the abdomen, sprinkled with powdered camphor.

Proclivity to Suicide.

The propensity which some persons experience to commit suicide, whenever the physical system is deranged and depressed in energy, is owing to the great *sensitiveness of the brain.* You can suffer or enjoy much, but the action of your brain is uneven. Did not your loved mother, before your birth, suffer suicidal emotions to disturb her spirit ? You should cultivate your organ of Hope, and absorb vitality from lower faculties of mind.

Poisoned by Mineral Acids.

Aquafortis (Nitric Acid,) *Marine Acid* (Muriatic,) or *Oil of Vitriol* (Sulphuric Acid,) if swallowed by mistake, may be antidoted by the abundant administration of *calcined magnesia,* or strong soap-suds, or saleratus, to neutralize the acid ; then give warm water to induce vomiting ; after which give plenty of flax-seed tea, or slippery-elm water, until the irritation has entirely subsided.

Sinking at the Pit of the Stomach.

Such stomach sensations are nervous, but pain in the right side indicates that the liver is diseased. REMEDY.—Take a tea-spoonful of powdered willow charcoal in a little cold water just before each dinner. Get some friendly hand to knead your stomach and side, as if to make bread of them, about thirty minutes after dinner.

Remedy for American Leprosy.

The best remedy for the American form of *Leprosy,* is an ointment of the following : Gum kino, half an ounce ; gum camphor, four ounces ; cajaput oil, two ounces ; mutton tallow, six ounces. Dissolve and mix over a hot fire. Use it when cold. Give yourself a thorough cleansing with soap and hot water. Afterward use this ointment on all parts of your person. It must be rubbed by the friction of your hand into your skin. Take a tea-spoonful of olive oil every morning for several weeks.

Dr. Valentine Mott, of the University Medical College, made lately, in one of his clinical lectures, a striking and novel statement. It was to this effect: That, to his mind, the conviction was irresistible that leprosy was the great progenitor of both syphilis and struma ; that they were all three essentially the same disease. His conviction, he stated, was founded upon extensive observations which he had been able to make upon leprosy in its various phases, while traveling in the East. The analogy between leprous and syphilitic sore throats and skin diseases, he instanced as being particularly striking and complete.

Rheumatism and Sore Eyes.

For chronic rheumatism in the joints, which are enlarged and drawn out of shape, there is nothing better than the following ointment: Flower of sulphur, one-half ounce; gum kino, one-half ounce; borax, one ounce; oil of amber, two ounces; turpentine pitch, one ounce; camphor gum, four ounces; mutton tallow, eight ounces. Melt and amalgamate over a slow fire, stirring the mixture steadily while dissolving; use it when perfectly cold, by rubbing it into the joints with all your strength and with all your Will. Always use your own hand, or get a friend to act for you. A piece of brown paper saturated with this ointment, and laid on your eyes at night, will do something toward giving them aid and comfort. Cheerfulness, and a mind of peace in the midst of discord, are important remedies for you.

Remedy for Tobacco Tremens.

A correspondent says that "it makes him almost crazy to leave off the use of tobacco. He longs to be free from the habit, but fears he cannot find himself strong enough to accomplish it."

We counsel all who find the habit so fixed, that to break it disables them for labor or business, to give up all attention to any occupation until the trembling nerves become steady. It is a kind of sickness, and the tobacco-chewer, in order to break up the narcotic habit, must lay up a few days, like any other sick person. Meantime, while the appetite is strong—perhaps ravenous and fickle—the WILL must forbid hearty eating. Let a man say: "*I will it,*" and his Will shall draw heavenly aid.

Boils Cured by Creosote.

Dr. Lynch (in the *Eclectic Medical Journal*,) in treating boils as a kindred disease to erysipelas, says: "In all cases, *creosote* is an *effectual local remedy*. It produces a blister, over which forms an eschar, or scab, when the sore readily heals. And I have never known a single failure, where the remedy has been applied *prior to the formation of a 'core,'* or the death of a portion of the areolar tissue. I have broken up whole crops of boils with this agent, without any other treatment. How it acts, or its *modus operandi* in these cases, let pathologists determine. But when the tumor has 'come to a head,' as a certain stage of its development, in common parlance, is termed, creosote will afford no service; and then suppuration should be favored by emollient applications, as poultices, fomentations, &c., till the 'core' is disengaged, when the ulcer rapidly heals under simple dressing."

Sympathetic Sore Eyes and Catarrh.

Your sore eyes may arise from sympathetic connection with a periodic disturbance in the head and nose. Your disease is, perhaps, one of the mucous membrane, commencing in your stomach, and terminating in catarrh and sore eyes. REMEDY.— Mix two ounces of sweet oil with half an ounce of camphor, over the fire. Rub this ointment into the skin of your stomach, in the cheeks, on the eyes, and very thoroughly manipulate it into your temples, and where the nose is most afflicted. Snuff sweet oil into your nose two or three times per day. Arise! Let blood flow into your feet and hands. Become very healthy, and, therefore, beautiful. Will it strongly. If the eyes are badly inflamed, and the eyelids very much thickened, then apply at night a poultice of rye flour mixed with the white of an egg. Always spread poultices for the eyes on linen cloths

Figs as a Cure for Cancer.

Mr. Thomas Puderton, an English gentleman, gives the following recipe for cancer, which he says has been of great service in several dangerous cases: Boil fine Turkey figs in new milk, which they will thicken; when they are tender, split and apply them, as warm as can be borne, to the part affected, whether broken or not. The part must then be washed, every time the poultice is changed, with some of the milk. Use a fresh poultice night and morning, and at least once during the day, and drink a quarter of a pint of the milk the figs are boiled in, twice in the twenty-four hours. If the stomach will bear it, this must be persevered in three or four months at least.

The efficacy of figs, in hastening the absorption of inflammatory particles in a cancerous sore, is indisputable.

THE REPRODUCTIVE ORGANISM.

Few readers permit themselves to live an hour in profitable conscientious meditation. No soul knows what Life is without occasionally coming into nearest rapport with all its holy essences and normal manifestations. We know that this age is heated with a boundless nerve-fire! Its subtile flame reaches and scorches almost every living soul. Thousands who would think calmly and act deliberately, cannot; they are disturbed every moment by the storm-beat of discordant circumstances. The elements are all astir. Therefore few persons stop to consider profoundly anything. The deeper your thought, the less likely is it to reach the multitude, for whom you think. Yet we would utter a brief word concerning the totality of the high powers of mind called "Wisdom."

It is the office of Wisdom to listen reverently to the sublime teachings of eternal principles. Wisdom sits on the glorious throne of all natural laws—presides over the innumerable manifestations of all constructive attributes—and is superior, both in position and power, to all the qualities and essences which fill the universe with moving organized bodies. Wisdom, therefore, is unconscious of either pride or prejudice. It contemplates and comprehends the teeming congregations of truths, which are perfectly displayed and infinitely expanded in every direction; and yet not for one moment does it (Wisdom) become self-conscious of its constitutional supremacy even to the truth itself. Therefore, the sublime simplicity of Wisdom is inexpressible—its order and form, its light and life, its grace and elegance, its uniform majesty, its constructive attributes, surpass the descriptive power of human language. Therefore let no earthly mind wonder that, even in the celestial realms of angel-existence, where exterior forms are known to their inmost centers, the revelations of Wisdom are pregnant with elementary simplicities, and with aphorisms congenial only to the intuitions of "the pure in heart."

Wisdom wants and seeks the Truth—independently and unconsciously of foregone conclusions. It is, consequently, a stranger to prejudice—is equally unmindful of prudential restraints—takes no knowledge of personal or private considerations—but, attracted alone by the heavenly breathings of the Divine and Eternal, Wisdom opens its myriad hearts to the influx of light from all points of the infinite radius, and meditates and legislates, unconscious of its sublime capacities and immortal powers, like a child of the Most High playing with the valley flowers, by the silently-flowing streams of the expanded earth, beneath the unfolded heavens.

But the office and characteristics of man's intellectual faculties are conspicuously different; for they are far more external than wisdom. If we gather together all the thinking faculties in man's nature, and systematically assemble all his judgmatic organs into one self-conscious congregation of thoughtful powers, it would then be most appropriate to class such an aggregation or congress of thinking faculties under one comprehensive term—"KNOWLEDGE." This name will express both the powers of acquiring information, and the acquisition also.

Under the term Knowledge we comprehend or include all such familiar names as "understanding," "reasoning faculties," "power of judgment," "intellectual organs," &c. It has long seemed to us that the most comprehensive name for the totality of man's self-conscious and external powers of mind is Knowledge; because Wisdom is interior and spiritual, or celestial, in its attributes, and because, in respect to office and characteristics, Knowledge is the natural opponent of what we have so imperfectly described as Wisdom in man's organization.

The inherent propensity, and, therefore, the office of Knowledge, is, to energetically and persistently make explorations. It is the embodiment of inquisitiveness, and the universal Yankee is, therefore, the embodiment of Knowledge. He is the latest revelation of the human family, and will go in pursuit of Knowledge " under difficulties " never dreamed of by ancient Greek or Roman. Art, Science, Mechanism, Commerce, are volunteer troops or body-servants—the express agents and incursive fillibusters — under the supervision and government of the one despot—Knowledge. Without faith, or fear, or politeness—but stately and lordly independent—it penetrates to the most secret recesses of human life. No temple is too holy for its presence—no subject too private or intricate for its investigations. It is bold, and audacious, and skeptical. Destitute of modesty or suavity, it "is bold to inquire" of everybody exactly what a body means to mean on every imaginable occasion. Knowledge, according to our definition, is rude, rugged, intrusive, invasive, introgressive, aggressive, and propulsively determined to ransack " the wide, wide world " in quest of gratification. It is constitutionally set to destroy every palace of mystery; to tame every wild horse; to conquer a peace in every land; to kindle inextinguishable chemical fires beneath every combination of unyielding elements; to become equal to the gods in comprehending good and evil; and, lastly, to annihilate, " both root and branch," every form of ignorance, error, and superstition.

Now it is useless to attempt to disguise the fact that this age is full of Knowledge-seeking propensities. The Wisdom-age has not yet dawned upon the glittering edifices of mankind. Only through a few delicate, courageous, inspired, retiring souls, have the rays of the eternal sun reflected upon the institutions of this era. This present age is strictly one of straightforward inquiry. But the dare-devil mode of such "inquiry" is often painful to the kind, prudential, and reverential. Man's external mind—his Knowledge department—says: "I want to *know*, to *understand*, to *weigh*, and *measure*, and *realize*—and what is more, I WILL." Moved energetically by this inborn determination to *know*, to be *convinced*, the young man, and not less the young woman, institutes the supposed necessary inquiries. If these be not satisfactorily answered by the parent, the friend, the playmate, or the teacher, then begins the ignoble and humiliating process of investigation—to wit, acquiring information of the secrets of life by means of intrusive conduct; by overhearing and eaves-dropping; by buying and stealthily reading of books on sexual physiology, written by libidinous authors for mercenary purposes; by closely watching the conduct of winged insects in the summer time; by observing the barn-yard fowls all the year around; by sensually cogitating upon the procreative habits of domestic animals; by waiting for the natal hour of, and curiously observing all the external facts connected with, the birth of lambs, colts, and calves—these, and a variety of other and less commendable methods of investigation, are instituted by the children of prudential, squeamish, and *unwise* parents and teachers, until Knowledge becomes to its possessor a source of incalculable mischief, instead of a glory and a blessing of endless duration. To attempt to prevent the ingress of Knowledge is folly of the shallowest kind. Knowledge is a bold highwayman, remember—a fillibuster, a sort of invisible Goth, or spiritual Vandal; it will not be delicate and graceful in approaching the penetralia of truth; for it is inherently stubborn, naturally hard-headed, peremptory, conceited, egotistic, despotic, resolved to waylay and obtain its rights at all hazards; and he is, therefore, inexpressibly unwise and inexcusably ignorant of the laws of mind, who attempts to set bounds to the penetrations of Knowledge.

Secret physiological knowledge, or information stolen out of circumstances, is a source of general mischief. If a young man imagines himself to be in possession of information concerning the reproductive organism, which is not known by young women, he is at once impressed with the unfortunate conviction that he is *superior* to his sisters, both by nature and attainment. It is an equal misfortune for the young woman to imagine that the essential and orderly functions of her organization are concealed from the world's comprehension. She is deluded by the popular faith that Nature has perfectly shielded her from the encroachments and incursions of scientific inquiry, and that it is impossible for her brothers and young masculine acquaintances to know anything positively of woman's procreative structure and functions. There are thousands of both sexes deeply impressed with this mistaken conviction. And yet, with consummate wisdom, the boundless Father and Mother have made it impossible that man should remain ignorant of woman, or woman of man.

But how inexpressibly essential it is that correct *Knowledge* should displace the present popular learned ignorance of mankind! Incorrect information, regarding the reproductive organism, is the fountain-source of much disease and misery. Boys, before their twelfth year, think they know all the *arcana* of human reproduction. Girls, between their twelfth and eighteenth year, indulge the same fancy. Both sexes assure themselves that they have acquired much secret information without the assistance of teachers, or the knowledge of parents; and each sex has also the conviction that the other knows something very different respecting the other's organization, functions, and experience; but the universal Mother hath wisely and lovingly built the temple of Life, and hath printed over the spacious vestibule, before the eyes of young and old alike, these words: "KNOW THYSELF." Hence universal ignorance is impossible.

We are impressed to do what we can to universalize as much of reproductive truth as we possess; believing that virtue and civilization effectuate from the interior realms of wisdom and causation. The centermost truth of man's position in the system of creation, is almost universally overlooked—viz.: that he (man and woman as one) occupies the highest place in the structural world. Mankind are the kings of organic structures. All forms, orders, degrees, forces, and principles, of the visible world, meet and co-operate in Man. He is an ocean composed of all the streams of life. The whole earth meets and mingles, and packs itself snugly in the human physical and spiritual bodies. His exterior is oft-times a rough casket—a thick, coarse, heavy, bad-smelling box, or trunk—but the priceless jewels and spotless robes are all within, waiting for the quickening power, for the congenial touch of the angel-hand, for the coming of the adequate commandment, saying: "Come forth and live."

In consequence of the universal overlooking of this centermost truth—concerning the position and microcosmical properties of mankind—a very great fallacy prevails respecting the facts and physiology of reproduction. Embryologists and investigators of procreative truths have made a mistake by simply regarding men and women as they do the males and females of any branch of the animal kingdom. Because woman has a mammalial organization, and because the elementary facts and incipient processes of reproduction are the same in mankind as in mammalial organisms in other parts of Nature, therefore, what is true of such animals is also true of mankind, and the secret of reproduction is no more profound in one department than in another. So the doctors reason among themselves.

Now the mistake, the mischievous fallacy, is this: Overlooking the sublime fact that a man and a woman are superior by organization, by endowment, by attainment, and by *Destiny*, to each and all the innumerable bodies and powers of the lower kingdoms, popular physiology contemplates generation in a rabbit with the same standard of judgment as procreation in the human family. The human species, quadrupeds, and *other* animals, are examined and disposed of in the same manner. Who would not *blush* before the looking-glass held up by such physiology? The females of quadrupeds, in nearly all books on reproduction, are the same as your sisters and your mother. The phenomena among human beings, of attraction, gratification, impregnation, gestation, parturition, and of nursing and fostering the young, are treated by physicians generally as though they were of no more importance than similar phenomena occurring in any family of the quadruped world. Constantly, therefore, mankind are compared with and treated like the animal creation. The superior is measured and adjudged by the *inferior;* and the effect is visible in the licentiousness and animalism of mankind. Young men are mis-educated by such popular physiology; and the young woman, from the same cause, knows not the real worth of her existence. The spiritual supremacy of *Human Life* is overlooked by many learned physiologists; and their students (the young physicians of the age) are not exempt from the degrading influence. Thousands of young human mothers have been ruthlessly treated by college-bred physiologists, simply because physical organs and reproductive functions in woman are supposed to be but little superior to the corresponding parts of animals.

Nearly all hypotheses on the subject of conception and reproduction, are nothing more than approximate truths. When the true philosophy of human origin is comprehended and taught, then will commence a *new era* in the history of earthly men and women ; then will dawn that divine faith which teaches, as through mythology so also by inspiration, that mankind were stationed "but little lower than the angels." It will one day come to be seen that woman inherits, by organization, not only all the physiological processes known as natural to the inferior kingdoms; but, what is infinitely more, that through her spiritual principles descend the impersonal essences of an immortal individuality. There is a divine life throbbing within these visible structures—a more important, a far more sublime *function* performed by the reproductive organism than is presented in modern books on anthropology. A human female is not comparable to the female of the quadruped world ; neither is a man to be measured and gauged by the male of any lower organism; because there is that superior *divinity* in the soul of each human being, which takes hold upon the amazing truths of eternity. We, therefore, hope to inspire woman with a higher and holier estimate of herself—so that, to her sisters, and daughters, and little children, she may impart the delicate lessons with the simplicities of Wisdom. Men regard their physical functions unworthily, because they have been mis-impressed, or because they have been left wholly in ignorance concerning the ulterior purposes of their physical existence and spiritual endowments. It is our impression that mankind, aided by the limpid light of physiology and pure principles, may and will attain to a higher mountain of truth.

A VOICE TO ALL WOMEN.

WRETCHED Mothers! Suffering Sisters! Unhappy Wives! we ask your most earnest attention. Our pen moves unfettered in your behalf. We write to you and *for* you, in perfect trust, as one loved and well-tried friend should write to another—" in freedom which the heart approves," and in confidence which friendship loves. We speak freely to you before all men, as it were in their very presence; and before all their sons also, appealing to every noble and saintly sentiment. Apologies are confessions of blunders committed, and graceful pardon-askings are acknowledgments of criminal deeds done. We are conscious of neither, and shall not, therefore, waste our moments in writing sentences of more than doubtful taste and courtesy. Our mission is to teach the golden ways of personal happiness, through obedience to Nature's immutable laws, which are, amid frozen oceans as on summer seas, the sacred will-decrees of eternal Father God. It is our inborn mission

>   ——" To show
> The secrets of the heart and mind ;
>   To drop the plummet-line below
>   Our common world of joy and woe,
> A more intense despair or brighter hope to find."

We do not, therefore, seek a quiet Eden-home, away from the world's intrusions, where the groan of the sufferer and the burdens of the broken-hearted cannot penetrate ; but instead, we take our seat near the throbbing core of the world's human life, by which we realize the prayers of paid priests and the pains of unpaid clod-hoppers by the road-side; the folly of fashion-rangers and the sorrows of wasted homes; all in vivid contrast with the wisdom of angel hosts and the grateful joys of the pure in heart.

At the center of this wondrous combination we gladly live, and move, and perform our mission. To this center come innumerable letters from the wealthy, the weary, and the weak —from the exceedingly poor—also, from the robbed, the spoiled, the hunted, the broken-down, and the very sorrowful. The bodily bonds and distresses of many mothers surpass the liveliest imagination. Thousands of young women, too, take their places in the legion army of invalids, all marching—slowly —sadly—steadily marching toward the insatiable cemetery, just behind yon trees or beneath the dark shadow of the village church. Not less than fifty descriptive letters, received from suffering women living in homes or huts all the way from the Atlantic to the Pacific coast, await an early answer—appealing tearfully for strength of health and the grace of harmony; and in almost all cases we observe a sort of *chronic ignorance* concerning physiological truths the most simple and important. In view of this condition among the young women and diseased mothers of our earth, we propose an unfettered utterance to the multitude. Let all men listen, for our discourse is to them not less.

HUMAN REPRODUCTION.—This term is derived from *reproduco*, meaning to produce again. It is the general term for that sacred function by which living, organized beings, reproduce their like. Anatomical details and physiological particulars are deemed unnecessary to correct government of such functions. But we consider a knowledge of the underlying principles absolutely essential to every human mind. Physicians do not presume to fully comprehend the inceptional phenomena of reproduction. Of all secondary processes and progressive transformations, however, the students of embryological science may be said to be familiar, and their definite knowledge in these respects has augmented the practitioner's power over the diseases of women.

But to affirm that physicians can cure the reproductive diseases of either men or women, is to assert what innumerable facts will hopelessly invalidate. Only a small proportion of American women are healthy. Almost all fashionable ladies are reproductively debilitated; they suffer periodically, and eighteen-twentieths are incapacitated for the divine office of reproduction. The working women in our farming counties, like the less industrious mothers of large towns and cities, are about equally diseased in the holiest functions of their being. All along the border regions of this continent, as upon the sugar, rice, and cotton plantations of the entire South, the women, "irrespective of age or color," are generally afflicted with uterine misplacements and prostrations exceedingly painful.

To teach the philosophy of prolification, in this connection, would be of little service. The finest memory would ere long forget the detail, and with the loss of memory would depart the salutary lessons. We will not write useless facts on this subject. Neither young nor matured mothers would make much progress by reading a learned description of *muscular cavities, hypogastric arteries, ganglionic nerves, uterine veins, fallopian tubes, ovaries*, &c., &c., because it is wisely and beautifully ordained, and it is so written in the Bible of Nature, which is God's only infallible revelation to mankind, that living organizations shall reproduce their like as it were without thought. It is an act to which both body and soul instinctively consecrate and unrestrainedly abandon the deepest vitality of their exist ence. Hence it will ever remain philosophically and theologically impossible to regulate the *act* of reproduction by intellectual statutes or scientific commandments. In truth, and to be plain, the reproductive office is exalted far beyond and above the stoical plane of intellectualism. It is Father God and Mother Nature in spontaneous conjunction, evolving, as from the unfathomable riches of their fountainous heart, the ascending forms of endless duration. What, then, shall be deemed the true standard by which to govern and regulate the process of reproduction? The only possible standard is a true knowledge of its principles, and a reverential regard for its sacred office. (We do not now speak of social and statute laws regulating the marriage relation, remember, but only *as a physician* of the reproductive functions in living organisms.) These principles are very simple and divinely beautiful. They consist of the highest and holiest proximity of exactly opposite embodiments, resulting in metempsychosis of mutually attractive forces, and eventuating in the complete organization of their inmost " image and likeness." Human offspring is formed for an immortal duration, and the parental vital bestowments are, in consequence, characteristic peculiarities of the spiritual body, during long periods after death, be the same *good, bad, or indifferent*. For it is very long before a living stream can rise higher than its vital source. We say all these things with the overflowing conviction that the people generally will receive some adequate conception of the almost eternal importance to be attached to the act of reproduction.

Reproductive Diseases.—Of all the hydra-headed forms and evils of *syphilitic* maladies we will not now write anything, reserving the sad and disgusting subject for a more suitable opportunity. But *all men* may expect a voice from us ere long, in behalf of the miserable and melancholy multitudes of every civilized country.

Let us mercifully and sympathetically roll up the curtain of feminine misfortunes. What are they termed? Their name is legion. The suffering sisters and the modest mothers cannot hide them from public observation. "Female Pills" are cunningly advertised in every city and country paper. "Uterine Tinctures" meet the eye of every child who stops to look at fancy articles displayed in drug-store windows, or within the show-cases against the wall. Medical charlatans everywhere sound the trumpet of quackery and pretension. They devote their entire genius and scientific experience to the treatment of "*Female Derangements.*" They portray the most distressing maladies, the most aggravating cases of "*Prolapsus Uteri,*" vaginal tumefactions, barrenness, suppressions, menstrual hemorrhages, &c., &c. Every disease of the female organism, is marvellously within the power of the mountebank's remarkable *pills, pastes, pessaries, and powders.* Merciful heavens! Holy angels of Light! Save and exalt our good and beautiful women! Shield them from the assaults of medical pretenders, and from the mal-practices of scientific vampires! Save them! When and by what method? Now! from this moment. By what means? By methods and practices hereafter to be specified. Of course, in these sweeping statements regarding "scientific vampires," we do not mean to reflect dishonorably upon any well-educated and gentlemanly physician.

In order to better appreciate woman's organal sufferings, we will glance at the many and various diseases to which, between the tenth and forty-eighth year, her reproductive constitution is more or less liable, under the potency and action of existing causes. Let us give them the hard names which they deserve: *Leucorrhœa, Fluor Albus, Dysmenorrhœa, Sterility, Menorrhagia, Uterine Hemorrhage, Amenorrhœa, Hysteria, Ovarian Dropsy, Uterine Neuralgia, Metritis, Ovaritis, Vaginitis, Hydrometra,*

Polypus Uteri, Cancer in the Womb, or Scirrhus, Physometra, Enlargement, Catalepsy, Epilepsy, Emaciation, and prostration of the general system.

These names are employed to convey an idea of most frightful and formidable maladies. But the pure English of them all is, that a great number of conditions and changes exist and occur, at different times, in the reproductive organism. A true philosophy of disease will classify these several and distinct affections as the *different forms and modifications of one derangement.* The original and primal disturbance is confined to the principal organ—namely, the uterus, or womb. This organ is subject to diverse misplacements. One is called "*Introversion,*" or the turning of the womb inside out; another "Oliquity," or a sidewise falling; another, "Retroversion" or a falling backward; another, "Anteversion," or a forward falling; all which, with still more particular modifications, are referable to one principal and primary cause—namely, to a prostration of the reproductive organism; and this effect never exists without a predetermining cause, which it is the moral duty of every mind to fully comprehend and promptly overcome.

CAUSES OF REPRODUCTIVE DISEASES.—It would seem that the most of woman's physiological sufferings are unavoidable. This appalling doctrine is inculcated, by our orthodox brethren, both from the popular pulpit and in all the literature over which they preside. They teach the theology of ancient India, of Egyptian darkness and bondage, that woman's menstruational diseases and child-bearing pains are the logical consequence of an " original sin." We read the Book of Nature with a different light beaming through its thought-laden pages. We discern that our Sisters need not suffer, and that our young Mothers are not called upon to pass the ordeal of twenty deaths in reproducing their like; but, otherwise, that our despoiled daughters and pain-haunted women may perform their sacred missions unscathed by disease, and unsullied by the animalism of man.

What mean these peace-destroying symptoms? Bearing down in the lower parts of the abdomen; heat, dull pain, burning weakness in small of the back; sore places on the spine; small of the back tender to the touch; dragging achings in the loins; indisposition for bodily exertions; dread of walking either far or fast; the feeling, now and then, of numbness or paralysis. Why are our married women so capricious of temper; so childish at times; so irascible; so given to transitions from heat to cold, from amiableness to peevishness and frettings; and why are so many afflicted with scrofulous swellings and ameness?

The causes cannot be disguised. Scientific secrecy is useless. Large words and gracefully turned periods may fulfill the ends of rhetoric or imposture, but the truth must be told, that *all reproductive diseases are caused originally by excessive and unrestrained indulgence of the animal inclinations.* Neither men nor women have comprehended the primal causes of their sufferings. Their children are born with broken-down blood-globules floating through their infant hearts. The vital prostrations of parents circulate, under the disguise of "scrofula," in the blood of their offspring. Their young bones absolutely *ache* with the voluptuous fatigue transmitted by ignorant parents. Eevry ganglionic center is a telegraphic station. It receives impressions and transmits the signs of disturbances from part to part. Who wonders that our children are *scrofulous*, and *rheumatic*, and fond of *stimulants*? Who, that can trace the relation between one cause and another, will grope around yet longer—asking the learned doctors to explain why women are sick and unfit for the ordinary duties of housekeeping? Or who, knowing the truth, will further practice the ungodly habits of intemperate reproduction ?

REMEDIES FOR REPRODUCTIVE DISEASES.—The principal immediate cause of woman's suffering is *Prolapsus Uteri*. From one side to the other of this vast continent—as we know by clairvoyant perception—the one great prevailing disease among women is *Falling of the Womb*. The organ is weakened by a variety of predisposing influences, and then falls in one or more of the several directions indicated in preceding sentences.

Now the treatment we propose is radical and positive. Of course you know that the paragon of all remedies, the faultless curative power within the pharmacy of every immortal mind, is *the* WILL. Let every Sister, whether daughter or mother, apply the *Pneumogastric principle!* Well or ill, diseased or healthy, do not fail to exercise this immaculate energy. Of the super-excellence of this remedy we need not further write. Suffice it to say that this discovery totally supersedes the unwieldy and lumbering medicines of the quack and the druggist.

In addition to the Will, which remedy many persons seem incapacitated for exerting to advantage, we further prescribe the *immediate disuse of coffee among all diseased women*. We can demonstrate that the action of coffee *is directly* prostrating to the reproductive energies. There are educated men and mothers who think (ignorantly enough!) that a young woman may be suffering with suppression of the menses, with Chlorosis, or Leucorrhœa, and yet not be afflicted with any misplacement of the uterus. This is impossible. The womb is invariably disturbed before any decided symptoms are developed. Coffee is the great enemy of woman, if habitually used as a part of her diet; for its positive effects are stamped upon the reproductive organs a few hours subsequent to each meal. All young women,

—whether ignorant or educated, whether married or not, whether American, European, African, bond or free—are hereby counseled to *drink no more coffee*. Of tea we need not here write, since the greater of the two evils, among our women, is the decoction of coffee. We can promise health to no woman, unless she immediately comply with this injunction.

Again: it will be impossible for our women to maintain healthy systems while their husbands insist on the *legal* rights of wedlock during the period of pregnancy. The custom is *devastating* to every feminine sensibility, to say nothing of the crippling and polluting effect exerted by it upon her reproductive functions. The expanded heavens are begemmed with eyes that see these matchless sins. Deadly, indeed, is the detestable effect on the sacred soul of woman. Let husbands and wives, and the friends of children, see to it; no conjugal relation should exist between the married during the period of uterogestation. Women can produce offspring with impunity, and without danger or severe pain, if their companions will but reverence and shield them. Let every true Brother do what is right.

> "Do what is Right, for the day dawn is breaking,
> Hailing a future of freedom and light;
> Angels above you are silent notes taking
> Of every action; then do what is Right."

Yes, do what is right! The scientific quack will pretend to you that his skill is adequate to the reparation of all damages occasioned by your disobedience. Beware! His pretensions are fatal to your spiritual progress. We can prove that physicians do not—because they cannot—cure the distressing diseases of women. Certain symptoms, indeed, they readily *modify, transpose, invert*, and *master* at length; but the fundamental, functional derangements of the reproductive system, no College of physicians can control. In fact, the honorable and the best educated physicians of the age confess themselves powerless in the presence of the radical diseases. Husbands, Brothers, Wives, Sisters! reflect on your sacred missions to each other, and henceforth honor your nature and its exalted destiny.

FEMALE MIDWIVES A NECESSITY.—The majority of married women are suffering from the misfortunes of mal-practice during confinement. Many sensitive natures cannot become reconciled to the universal custom of masculine assistance. Heaven grant that all women will very soon openly remonstrate against the shocking intervention. Beautiful souls always shrink from the wretched system of "doctoring" a child into existence. What are your intelligent wives and daughters doing? Able physicians should have wives, or daughters, or agreeable female associates, who comprehend the facts of pregnancy, and all the *modus operandi* of parturition. Hundreds of beautiful women, in every station of life, are suffering from unskillful treatment at the hands of man. The forcible disengagement of the *placenta* has disabled many a noble lady for life. All these evils are preventible, and they should be preached against and abandoned by every harmonial soul.

INTUITIVE GLIMPSES OF TRUTH.

The human spirit is framed for the perception and enjoyment of heavenly realities. Its strongest passion is the development and possession of the Beautiful. Thousands would prefer physical Beauty to spiritual happiness. Beauty is more attractive and influential, in the soul's juvenile periods, than intellect or moral excellence. The best feelings of the aged are moved by " the *beauty* of holiness."

All young dreams are mingled with the elements of imperishable beauty. The pictures of hope and the longings of the imagination are painted on the canvas of eternity. The soul swells with unutterable yearnings for the speedy fulfillment of the prayer : "Thy kingdom come—on earth—as in heaven." And why ? Because the very elements of which the spirit is constituted are the exclusive property of the Summer World to come. The next sphere is imagined as a possible part of this, the rudimental. Wherefore men invest this primary existence with those imperishable characteristics which, in the very constitution of the universe, are inseparable from the Spirit Land.

The soul's inmost fount of intuition is ever and anon touched by the presence of an angel. It throbs with the impulsions of immortal hope. It swells with emotions of sublime discontent. The miseries and materiality of this life become clogs, evils, and burdens, too hard and too unnecessary to be longer borne. The impatient spirit prays "for the kingdom of Heaven on earth."

Human association is no longer satisfying. The soul goeth forth thirsting and hungering after righteousness. From the invisible Spirit of Creation, the God of Nature, the soul obtains rest and hope. The frowns of Winter are exchanged for the sweet smiles of Spring. The abundance, the dignity, the beauty and perfections of the Spirit Land, are visible in Nature's birds and trees, in her landscapes and flowing streams. Dreams of deathless enjoyments, visions of universal happiness, are revived like flowers in the summer time, and every effort of the fancy is toward the Beautiful and the Immortal.

The human spirit is longing—through its hopes, its fancies, its prayers, its reasonings, and its fears—for its native eternity. The most vitiated mind is accessible through some one of its many beautiful avenues. The practical voluptuary, the soul that is wandering through the wilderness of trespasses and sins, is susceptible to the magnetism of good and truth. Sympathy, or music, a holy picture, poetry, intellect, or beauty—each soul is approachable through one or more of these ; and he who thus reaches the fallen, or sorrowing, and continues the labor of elevation all around the foundations of the unhappy outcast, will save the sufferer with an everlasting salvation. Who wants a richer reward ?

The power of the spirit to imagine and anticipate the realities of the Spirit Land is so perfect, that, on its arrival there, a feeling of familiarity steals over the mind, as though it had many times before witnessed the same scenes. Poets, and painters of landscapes, color all their best thoughts with the vague tints of immortal beauty. When such minds arrive at the eternal world, and open their spiritual senses upon the verities and landscapes of the Summer Land, they feel instantly *at home* and content, having anticipated many of the external forms of truth and beauty which characterize that existence.

All love of the Beautiful, all passion for Music, all attractions of Sympathy, all dreams of Love, all aspirations for Wisdom, all yearning after Purity, all prayers to know and possess the whole Truth—all these are good, for they attest the immortal existence in store for the spirit, to which, by virtue of its origin and essence, it is indissolubly allied.

The most physically wretched, the morally insane and crippled, the intemperate, and brutal, and vicious, the saint and sinner—all, as one man, cry naturally to Heaven! The soul instinctively dreads death, because it cannot die: but confounding the physical fact with the spiritual conception of it, the tongue prays for the speedy advent of the kingdom of Heaven. The memory gathers the garlands of intuition, revives the vague realities of past dreams of truths, and impresses the intellect with a sense of earthly possibilities. Wordsworth gave the internal workings, thus:

> ——"On the mind
> They lay like images, and seem almost
> To haunt the bodily senses."

In this manner the spirit's intuitions overflow the sensorium, so to speak, and flash before the inward faculties with all the tangibility of a scene of yesterday; so that, in the lapse of years, the intellect can neither discriminate nor separate the pictures or dreams of intuition from impressions derived by actual contact through the senses with the landscapes and objects of this present world.

Wherefore thousands of sincere natures, not knowing that the mind is a native of eternity, imagine that God designs supernaturally to extemporize a celestial kingdom of eternal bliss, displacing the earthliness and perishableness and misery of this sphere with a "new heaven and a new earth, wherein dwelleth righteousness." Such pray for the transformation of earth into a paradise. As well might the unfolding flower supplicate its solar god, the sun, to come down from the dizzy hight and burn its glory into the earth's cold bosom. The soul yearns toward the remote sphere just as the rose looks lovingly up into the great, beautiful eyes of the sun. But like the yearning rose, the human soul, while gathering experiences in the rudimental sphere, is blest by enkindling rays descending from remote orbs, and would not be benefited by immediate proximity to the source of every blessing.

To the perfectly healthy-minded there is, properly speaking, nothing of what is called "imagination." Every vague intuition is a foreshadowed positive fact; each internal conception is interwoven with truth; every dream will work out a fulfillment; and the reward of merit even goes further, transforming the commonest pleasures into spiritual happiness. But the discordant, the ignorant and diseased, on the other hand, are beset with penalties and punishments of every shade and magnitude. They will turn from the most comfortable circumstances into sufferings and misfortunes purely imaginative in origin. Portentous omens and silly prognostics infest diseased and discordant minds, while the inspirations of peaceful natures are filled with the golden verities of eternity.

Bright and beautiful are the countenances of the pure in heart. Surpassingly beautiful and bright are the eyes of the higher angels. The meditations of harmonious intelligences are celestial and heavenly.

But there is a beauty even in the *external* form of every created thing. There is an expression of tender love and saving wisdom in the eyes of guardian spirits. They bring beauty and light in their garments. How pure—surpassingly pure—are they, in all their deeds of mercy among the children of men! With what graceful tenderness do they bend over the fallen in battle! They ride upon the bosom of the rivers of magnetic fire—from their lovely homes in the Summer Land, through the star-paved immensity, down to the couch of the sick and dying. Into the undulating brain of the sick one they breathe the breath of celestial love. They mitigate his pains by the aroma of their hearts, and they impart beautiful dreams of coming happiness to the departing.

How incessantly employed in deeds of friendship are all the noble and pure in the Summer Land! There are no latitudes, no longitudes, to the inhabitants of celestial spheres. They leave their valleys and plains for the mountains and rough places of earth. They depart from their beautiful gardens, and from the enjoyment of their luxuriant possessions, to mingle their feelings with those on earth who pray and *work* for the reign of Freedom. Goodness infinite! is proclaimed by the noblest tongues. The atmosphere of a deathless divinity fills all the space in which they dwell. They would spread such a gospel and exhale such aromas throughout all the habitations of men. The fallen woman, the lost child, the dying soldier, is covered by the temple of their love. The sky is full of bright eyes, and the earth is peopled with dark objects, and those eyes and those objects meet both day and night. Each object is instinct with claims upon the heavenly visitors, and each visitor is prompted to perform some kindly office for the sake of humanity.

Garments of whiteness are thrown around the fallen spirit, and a magnetic life steals into the heart's darkest chambers. Whether a human being ascends from the battle-field, or from the retirement of the bed-room couch, the heavenly visitors descend to enfold him in their exalting love. There are hundreds of philanthropic celestial visitors to every village; and there are ten helpers for every man whose spirit is dislodged by accident and misfortune. Think how many tender-hearted mothers dwell in the Summer Land! Of the benevolent and unselfish, who once lived on earth, there are millions in the adjoining world. A stream of constant philanthropy flows from them earthward; and, when possible, they lift the down-trodden, and save the falling from a lower depth.

In all this we behold the face of Mother Nature, and feel the omnipotent hand of our Father God. The manifestations of God are goodness, and truth, and wisdom; but the love, and purity, and philanthropy of the world, are from the heart of Nature. The attributes of humanity are more displayed in the Spirit Land. This world is rudimental—is filled with ignorance, selfishness, and strife.

All men in *this* world are naughty children; in *that* sphere all children are beautiful and good. The selfishness of earth is not fostered in the Summer Land. Philanthropy, not hatred, arches the door of every heart. Lovelier and lovelier beam the countenances of the guardian hosts; and sweeter than the waters of the rivers of Paradise, is the breath of every one who visits mankind on missions of mercy. What a Moral Police are the strong soldiers of the higher worlds! What beautiful "Sisters of Mercy" are the lovely nurses of the heavenly hospitals! How free from earth's drugs are all their medicines for the sick and earth-worn traveler! Bright jewels of truth adorn the crown of every philanthropic spirit. And lovelier than the multiplied flowers of a thousand summers, are the faces of those who still the troubled waters of earth!

Why do they not lift men's bodies and minds above the sphere of Disease and War? Why do they not extinguish the flame of passion? Why not put forth all their combined powers to heal physical and political sores? Will they not, having the power and the wisdom, save mankind from destroying one another?

Beautiful questions! All answered by the fact that this would is the *rudimental sphere of human existence.* Millions of angels, all obedient to the laws of the infinite Good, cannot do impossible things. Thunder will reverberate in that world which is filled with lightning. Disease and War will disturb all men who foster the fires of passion. Passion is the electricity of the mind, and war is the thunderbolt. Ignorance is the diabolical monster of the human mind, and selfishness is the " roaring lion " that goes up and down the earth, seeking whom it may " devour." Progression is the angel of our deliverance, and our heavenly visitors can but hasten the day of its power. The race, like the globe, revolves. Its revolutions are less and less eccentric, as the wheels of time fly swiftly round, but the perfect *circle* is not yet reached. We must work out " our own salvation " from the causes of unhappiness. The angels will help us just in proportion as we help ourselves

THE END.

CONTENTS.

	PAGE
Automatic Forces,	139
Animal Food, Uses of,	176
Antiquity of the Bath,	182
Arsenic Eaters of Styria,	204
Aphtha Chronica, Cause and Remedy,	286
Abdomen, Swelling of,	293
Asthma, Chronic, Treatment for,	336
Air Passages, Irritation of,	337
Asthma, the Wheezing or Hay,	340
Brain Life and Lung Life,	67
Blood, Bile, and Bowels,	81
Body or Spirit, Which is First,	183
Blood, Incorrigibleness of,	190
Bite of a Rattlesnake,	210
Blood, the Wonders of,	221
Bad Dreams Every Night,	255
Brain Rest Absolutely Necessary,	263
Brain Fits, Cause and Remedy,	268
Brain Fits and Incipient Epilepsy,	266
Brain, Softening of,	258
Biliousness, Remedy for,	288
Bowels, Looseness of,	291
Bowels, Heat and Pain in,	291
Bilious Vomiting, Remedy for,	293
Bilious Medicine,	295
Beware of the Tyrant, Disease,	309
Bronchial Weakness, and Neuralgia,	320
Bronchitis, and Mucous in the Throat,	321
Bleeding from the Nose,	332
Blood, Wasting of,	337
Blisters, Small White, on Children,	355
Burns, Rapid Cure for,	369
Burns, Excellent Mixture for,	370
Boils Cured by Creosote,	394
Clairvoyance, Medical Value of,	25
Clairvoyant State,	94

422 CONTENTS.

	PAGE.
Cause and Cure of Impatience,	147
Calomel, the Use of,	200
Change of Clothing in Spring,	202
Cod Liver Oil and Cocoanut Oil, Influence of,	205
Corn-fields and Magnetism,	213
Corruption and Groans,	219
Cotton for Garments,	220
Cold Feet, a Cure for,	220
Caging of Birds,	227
Chewing, the Disgusting Habit of,	234
Coup de Soleil, or Sun-stroke,	252
Confidence in Mother Nature,	257
Cords of the Leg Deficient, Remedy for,	271
Costiveness, Cure for,	295
Canker on the Tonsils,	316
Croup, Parasitical, Cure for,	326
Cold-taking, Prevention of,	330
Coldness in the Back of the Neck,	331
Cough, and Incipient Consumption,	335
Catarrh and Cough, Treatment for,	340
Cephalic Pills, the Use of,	344
Cold in the Head, Cure for,	344
Cultivation of the Sense of Smell,	345
Canker Humor, Medicine for,	351
Chapped Face, Hands, and Lips, Cure for,	357
Cancer Cured by Nitro-muriatic Acid,	359
Cancer Doctor, Confessions of a,	359
Cure for Foul Air,	373
Cholera Morbus and Cholera Flatulenta,	381
Catarrh and Difficult Breathing,	390
Cancer, Figs as a Cure for,	395
Disordered Liver, Symptoms of,	135
Duodenum, Inflammation of,	136
Do Infants Grow in Heaven?	191
Dreams in Disease,	216
Duty of the Skin,	229
Don't Stand on the Track,	232
Drunkenness, an English Cure for,	238
Dyspepsia and Despair,	290
Dyspepsia, Remedy,	293
Discordant Spirits and Spiritual Afflictions,	308
Dysentery, or Bloody Discharge,	310
Dysentery, Treatment and Cure,	311
Dysentery, Chronic, Treatment for,	312
Dyspeptic Pain about the Heart,	316
Diptheria and Sore Throat,	317
Difficult Breathing, Remedy for,	330
Duodenum and Sore Eyes,	338
Deafness, Sulphuric Ether for,	341
Deafness, Common, Cure for,	346
Drink for the Sick,	369
Effect of Air on Mind,	68
Exhausted Primates in Man,	155

CONTENTS.   423

	PAGE.
Electricity of Immensity,	158
Early Rising Triumphant,	164
Effects of Tea on the Body,	207
Electricity in Vital Processes,	211
Enlarge your Thoughts and Perceptions,	231
Electricity and Phosphorus in Animals,	242
Earth's Polarities at Night,	255
Epileptic Fits, Treatment for,	264
Elixir of Life,	285
Evils of Eating for Amusement,	295
Effect of Disease on Life,	343
Ear, Ulceration of,	347
Erysipelas, Chronic, Treatment for,	350
Eruption, Intermittent,	351
Eruption, and Juvenile Excitability,	353
Eruption, Prurigo Pruritis, Cure for,	354
Eruption, Pityriasis, Red and Rough,	356
Eye, Diseases of the,	362
Eyes, Chronic Sore and Weak,	364
Eyes, Dimness of Sight of,	364
Eyes, Ophthalmia, or Inflammation of,	365
Eyes, Paralysis and Total Blindness of,	366
Eyes, Obliquity of Vision of,	366
Eyes, Spots, Webs, &c., before the,	366
Eyelids, White Spots on,	367
Eyes, Sympathetic Sore, Wash for,	367
Emetic, Results of,	371
Fable of Vampires, Origin of,	174
Food as a Medicine,	176
Fashionable Languor, Remedy for,	198
Frost-bitten Persons, Treatment of,	205
Fits of Indigestion,	265
Failing Memory, Remedy for,	283
Fluor Albus, Cure for,	304
Fever Sores and Tumors,	348
Freckles, Cause and Cure of,	356
Fevers, Origin and Philosophy of,	361
Fever, Continued and Exhaustive,	361
Fever and Ague, Remedies for,	386
Female Midwives a Necessity,	413
General Instructions to Invalids,	37
Gastric Methods Explained,	75
Gluttony, Evils of,	178
Gen. Debility, No Compromise with,	196
Grapes, Curative Properties of,	215
Gall and Spurzheim's Works,	261
Goitre, or Big Neck, Remedy for,	328
Hints to the Debilitated,	34
How to do Good,	115
How to Balance the Human System,	159
Healthy Food, Preparation of,	172
Habits and their Consequences,	195
Hahnemann's Materia Medica,	197

CONTENTS.

	PAGE.
How to Live One Hundred Years,	199
How to Keep the Teeth Clean,	201
How the Roots of Plants Feed,	213
How to Quiet Children,	232
Hemorrhoidal Infirmities, Causes of,	243
Hydrophobia, Cause and Cure of,	249
How the Will Acts on Nerve-centers,	259
Head Hot and Feverish, Remedy,	275
Headache—Periostitis,	279
How to Cast Out the Devil,	289
Honey and other Sweets,	299
Heart Disease—Hypertrophy,	313
Heart Disease, Remedy for,	314
Heart, Palpitation of,	315
Heart Disturbances and Nervousness,	315
Hyper-oxygenation, Theory of,	333
Hair, Removal of Superfluous,	374
Hair, Restoration of,	374
Hair Tonic, Recipe for a,	375
How to Retain a Good Face,	381
Individual Responsibility in Overcoming Evils,	109
It is a Pleasant Thing to Die,	121
Iron Magnets, How to Make,	208
Indian Spirits, the Influence of,	226
Infusorial Reproduction,	230
Intemperance, the Madness of,	235
Inferior Desires, Cure for,	303
Ill Temper in Children, Cure for,	343
Itch, Use of Sulphur for,	352
Itch and Tetter, Cure for,	356
Intuitive Glimpses of Truth,	414
Joints, Pain in,	342
Kidneys, How to Stimulate,	297
Kidneys and Bladder, Weakness of,	304
Leading Pathological Propositions,	127
Longevity, Secret of,	179
Lockjaw, Remedy for,	275
Liver, Derangement of,	286
Liver, Torpid State of,	287
Lumbago, Treatment for,	307
Liver, Tuberculation and Adhesion of,	325
Lungs, Negative Condition of,	334
Lung Fever, After-effects of,	335
Lungs, Hemorrhage of,	340
Lungs, Soreness in,	341
Lightning, Persons Stricken by,	373
Leprosy, American, Remedy for,	392
Man's Recuperative Power,	31
Morality of Pure Air,	77
Magnetism, Human, Philosophy of,	87
Magnetism, Source of,	89
Magnetic Polarity,	90
Magnetism, Practical Experiments in	92

CONTENTS. 425

	PAGE.
Magnetism as a Medicine,	97
Magnetic Processes,	101
Magnetism, Cautions for Operators,	105
Magnetic Treatment of Insanity,	106
Mind and Matter—their Relations,	120
Man's Telegraphic Power,	140
Magnetic Disturbances in the Atmosphere,	181
Mothers, a Word to,	190
Milk and Water as Beverages,	221
Minerals in Vegetation,	222
Man in the Animal State,	222
Medicine for Every Home,	223
Magnetizers, a Word to,	230
Man's Voluntary Powers,	239
Mind in Sleep,	241
Man who can Will,	260
Multitude of Sins, Remedy for,	281
Morphia, How to Overcome,	283
My Dyspepsia and my God,	290
Milk, Presence of,	302
Menstruation, Painful, Cure for,	306
Marrow, the Source of,	307
Magnetic Treatment for Intoxication,	308
Melancholy and Meanness, Cure for,	343
Mouth, How to Wash the,	375
No Infallible Remedies,	30
Nature's Progressive Energies,	143
Natural and Artistic Beverages,	239
Neuralgia, or Nerve-pain, Philosophy of,	267
Neuralgia, Drug Treatment for,	269
Neuralgia, Sudden, Remedy for,	270
Nervous Burnings and Pain,	278
Nervous Debility, Treatment for,	284
Nervous Trembling at the Stomach,	307
Nervous Rash,	354
Nutritious Food for the Sick,	368
Nervous and Convulsive Diseases,	380
Origin of Disease,	19
Od-Force and the Odoscope,	149
Orange Before Breakfast,	163
Overdosing, the Evil of,	207
Opium Eating, Penalties of,	236
Origin of Physical Beauty,	255
Objections to Deep Breathing,	329
Origin and Use of the Probang,	342
Odor from the Nostrils,	345
Old Sores, Treatment of,	348
Onions and Cider as Medicines,	376
Origin of the Human Spine,	377
Pearly Gates of Science,	9
Philosophy of Disease,	17
Periodicity of Disease,	20
Pneumogastrical Discovery,	45

36*

CONTENTS.

	PAGE.
Pneumogastric Nerve, Position and Function of,	49
Physiological Virtue,	62
Purifying Ordeal,	76
Psychological State,	93
Physics and Metaphysics,	150
Physical Strength and Energy,	168
Power-generating Habits,	170
Pseudo Health,	183
Principle, the Meaning of,	188
Positive and Negative, Definition of the Terms,	212
Patience, as a Medicine,	217
Pleasures of Home,	224
Pure Alcohol as a Medicine,	233
Physical Evils are Transient,	239
Piles, Remedy for,	245
Pneumogastric Remedy, Passive Use of,	270
Purpose in Pain,	271
Pain in the Face and Neck, Remedy for,	271
Pain in the Neck of Housekeepers,	272
Pain in the Joints, Remedy for,	272
Pain Between the Shoulders,	273
Pain in the Right Side,	274
Pain in the Breast,	274
Painless Paralysis, Treatment for,	280
Prepared Female Organism,	301
Pulmonary Weakness and Irritation,	322
Poisoned Wounds, Treatment for,	370
Poisoning by Phosphorus, Remedy for,	370
Poisoned by Saltpeter,	372
Poisoned by Ash-leachings,	373
Poison, Bee or Serpent,	373
Palm Blister as a Remedy for Disease,	377
Putrefaction, an Antidote for,	379
Physical and Spiritual Man,	383
Physiology of Courage,	384
Phthisic, or Asthma,	390
Pin Worms in Ano,	390
Poisoned by Mineral Acids,	391
Rheumatism and a Bad State of the Liver,	65
Relation of Lungs to Brain,	73
Rain and Electricity,	214
Red and Black Pepper,	220
Remain in your own Climate,	258
Rheumatism, Membranous, Remedy for,	276
Rheumatism, Periodical, Remedy for,	277
Resignation as a Medicine,	282
Remedial Use of Sugar,	299
Ringworm, Cause and Cure of,	352
Rheumatism and Sore Eyes,	393
Reproductive Organism,	396
Reproduction, Human,	405
Reproductive Diseases, Causes of,	409
Reproductive Diseases, Remedies for,	411

CONTENTS.

	PAGE.
Self-healing Energies Better than Medicines,	40
Sympathetic Ganglia, Pathological Offices of,	55
Singular Physiological Facts,	70
Structure of the Stomach,	72
Spring Time Diseases,	127
Spring Beverage,	130
Skin Diseases, Origin of,	131
Spiritual Briers and Thorns,	145
Superior Condition,	96
Self-Rectification,	187
Suicide, the Evil of,	197
Sweet Oil in Relation to Poison,	201
Surgery, American Improvements in,	209
Sleep, without Dreaming,	216
Summer Foods and Drinks,	218
Silk Dresses, Effect of Wearing,	218
Sympathy with an Amputated Limb,	225
State of the First Man,	227
Spiritualistic Superstition,	227
Strong Drink, Substitute for,	236
Sleep, Proper Amount of,	253
Sleeplessness—its Cause and Cure,	253
Spasms, Periodical,	270
Stiff Ankle, Remedy for,	276
Sick Headache, Cure for,	292
Sour Stomach, Cure for,	299
Sweating of the Extremities,	300
Spina Bifida, Cure for,	306
Scarlet Fever, Prevention of,	339
St. Vitus' Dance, Remedy for,	342
Scrofula and Erysipelas, Medicine for,	349
Scar on the Face,	358
Scarlet Fever. Nature and Cure of,	360
Strychnine, Antidote for,	371
Swallowing Lizards, Frogs, and Toads,	372
Sweet Oil as a Remedy for Poison,	385
Stomach, Tenderness at the Pit of,	391
Suicide, Proclivity to,	391
Stomach, Sinking at the Pit of,	392
Sore Eyes, Sympathetic, and Catarrh,	394
Treatment of Disease,	22
Temperaments, Marriages of,	153
Tea, Coffee, Alcohol, and Tobacco,	184
Temporary Marriages,	193
Thoughts of a few Good Heads.	215
Twenty-one Systems in the Human Body,	223
True and False Hospitality,	228
Traveling as a Medicine,	229
Temporary Insanity, Forms of,	262
Tone of the Stomach Destroyed,	296
Throat, for Lump in,	316
Throat, Irritation in,	322
Throat, Swelling of,	324

CONTENTS.

	PAGE.
Throat, Malignant Sore, Remedy for,	318
Throat, Sore, and Bitter Stomach,	319
Throat, Consumptive Irritation in,	320
Throat, Sore, and Naked Arms,	323
Throat, Clergyman's Sore,	324
Tuberculosis, Remedy for,	328
Throat, Enlargement of Glands in,	329
Teeth, Loose, and Decaying Gums,	347
Thyroid Gland, Enlargement of,	347
Thin Shoes and Wet Feet,	376
Typhoid Fever, Treatment for,	382
Tea, Coffee, Cocoa, and Chocolate,	385
Tobacco Tremens, Remedy for,	393
Unity of the Universe,	193
Unhealthy Occupations,	203
Unequal Bodily Development,	256
Urinary Weakness, Remedy for,	296
Uterus, Displaced,	306
Vital Electricity, Action of,	74
Vegetarianism among Animals,	230
Value of Sunlight in Houses,	231
Vitality, Depression of,	285
Vaginitis, or Irritation, Cure for,	305
Voice, Loss of, Remedy for,	327
Virus Developed by a Scratch,	355
Will Power—How to Exercise it,	52
Will Power—Additional Instructions,	138
Why Primary Processes of Formation are Discontinued,	192
What is Evil?	194
Where goeth the Soul of Things?	194
Wretchedness at Home,	224
Would you Stop the Flowing River?	229
What and Where is Heaven?	232
Will-energy at Work,	259
What the Will can Do,	261
Weakness and Emaciation, Remedy for,	279
Weakness and Pain,	280
Water in the Stomach,	294
Weak Stomach, Medicine for,	298
Worms in Children, Remedy for,	300
Worms Under the Skin,	357
Wounds and Injuries, Poultices for,	369
Water Dressings for Wounds,	369
Water, the Medical Uses of,	380
Women, a Voice to All,	404

LIST OF THE WORKS
OF
ANDREW JACKSON DAVIS,
IN THE ORDER OF THEIR PUBLICATION.

	PRICE.	POST.
1. NATURE'S DIVINE REVELATIONS.	$3 75	50
2. A CHART. (On Rollers — by express). . . .	2 00	
3. THE PHILOSOPHY OF SPECIAL PROVIDENCES. .	0 20	02
4. THE GREAT HARMONIA. Vol. I. — The Physician.	1 50	20
5. THE GREAT HARMONIA. Vol. II. — The Teacher.	1 50	20
6. THE PHILOSOPHY OF SPIRITUAL INTERCOURSE. Paper, 60 cts. Cloth.	1 00	16
7. THE GREAT HARMONIA. Vol. III. — The Seer. .	1 50	20
8. THE APPROACHING CRISIS.	0 75	08
9. THE HARMONIAL MAN. Paper 40 cts. Cloth. .	0 75	12
10. THE PRESENT AGE, AND INNER LIFE.	2 00	24
11. FREE THOUGHTS CONCERNING RELIGION. . .	0 20	02
12. THE GREAT HARMONIA. Vol. IV. — The Reformer.	1 50	20
13. THE PENETRALIA.	1 75	24
14. THE MAGIC STAFF. An Autobiography. . . .	1 75	24
15. THE HISTORY AND PHILOSOPHY OF EVIL. Paper, 40 cts. Cloth.	0 75	12
16. THE GREAT HARMONIA. Vol. V. — The Thinker.	1 50	20
17. THE HARBINGER OF HEALTH.	1 50	20
18. ANSWERS TO EVER-RECURRING QUESTIONS. . .	1 50	20
19. MORNING LECTURES.	1 75	20
20. MANUAL FOR CHILDREN'S LYCEUMS. Cloth, 80 cts. Gilt and Leather.	1 00	08
21. DEATH AND THE AFTER-LIFE. Paper, 35 cts. Cloth.	0 60	08
22. LYCEUM MANUAL. Abridged.	0 45	04

A discount of 25 per cent. on the above retail price when the whole list is purchased.

Sent by express (C. O. D.) as soon as ordered.

☞ Persons ordering books sent by mail should add the amount mentioned for postage.

Address,

BELA MARSH, PUBLISHER,
14 Bromfield Street, Boston.

www.ingramcontent.com/pod-product-compliance
Lightning Source LLC
Chambersburg PA
CBHW020739020526
44115CB00030B/612